T0304614

Nanomaterials for Chemical Sensors and Biotechnology

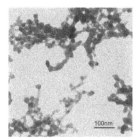
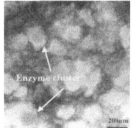

Nanomaterials for Chemical Sensors and Biotechnology

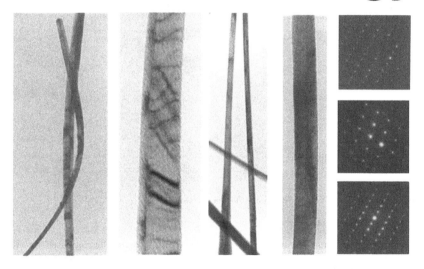

Pelagia-Irene (Perena) Gouma

State University of New York, USA

CRC Press
Taylor & Francis Group
Boca Raton London New York

CRC Press is an imprint of the
Taylor & Francis Group, an **informa** business

CRC Press
Taylor & Francis Group
6000 Broken Sound Parkway NW, Suite 300
Boca Raton, FL 33487-2742

© 2010 by Taylor & Francis Group, LLC
CRC Press is an imprint of Taylor & Francis Group, an Informa business

Visit the Taylor & Francis Web site at
http://www.taylorandfrancis.com

and the CRC Press Web site at
http://www.crcpress.com

*This book is dedicated to the loving memory of my
parents, Georgios and Ekaterini Gouma. They are dearly missed...*

Preface

Upon embarking on an independent research career one faces the challenge of choosing a topic within his/her field to study in depth so as to make new and significant contributions to science, engineering or medicine. The authors' earlier involvement with the characterization of colloidal 3D networks of metal oxide (titania) nanoparticles and their thin films had suggested that these exhibit catalytic properties far superior to those of their micro-sized equivalents. Heating the nanostructures at temperatures that were considered "safe" for larger particles, the nano-entities would undergo a phase transformation and coarsen abnormally (due to oriented attachment) losing their nanostructure advantage. Studying the gas sensing properties of ceramic nanomaterials, was the beginning of a learning path to the wonders of nanotechnology.

Sol-gel 3D nanomaterials led to 2D nanobelts and from there to 1D nanowires that can detect just a few molecules of a specific gas with high selectivity. Nanomaterials synthesis methods from the conventional: electrodeposition, physical or chemical vapor deposition (PVD, CVD) to the exotic: laser ablation, flame spray synthesis, and the latest nanomanufacturing craze: electrospinning, have all been used in the authors' research and are discussed in this text. Ceramic materials were combined with biomolecules to make novel bio-doped composites and resistive biosensing platforms for the rapid screening/determination of pathogens and biothreats; and electroactive polymer transducers were introduced for enabling ion to electron transport. Nanofiber mats became the ultimate "active" and "smart" structures for bio-chemo-sensing and actuating, ultimately leading to 3D scaffolding architectures for implants.

So, this is the journey that you are invited along through the contents of this book. The hope is that you'll find the adventure as thrilling and fulfilling as the author has felt it (and she is still finding it) to be, with a strong belief that nanotechnology has the potential to lead to breakthrough materials and devices that will change our lives immensely for the better. Non-invasive diagnostics, personalized and telemetric medicine applications, skin, bone, and organ regeneration, ubiquitous environmental monitoring, novel energy sources, are a few of the promises that are becoming realized with time. It is with great anticipation that the impact of the advances described in this book on the human life and welfare in our lifetime is envisioned.

This book involves some part of each of the sciences, engineering and medicine fields: nanomaterials, nanotechnology for chemical /bio-chemical sensing, and nanomedicine, i.e. the application of nanomaterials in biotechnology. The fundamental principle that has enabled the recent progress in each of these fields is that materials at the nanoscale (sized between 1–100 nm) behave in a different manner than their bulk form, exhibiting size dependent properties. It has only been in the last 10–15 years that this notion was introduced to the research communities and it has still to catch up with the general population. There are graduate college students who have never heard of nanomaterials even today. There is a layman population who's read science fiction novels about nano-robots and other scary products of nanotechnology, feeling confused about this new science. And there is a real need for advanced technologies to protect the human health and welfare as well as the environment.

The above became the stimulus that led the author to introduce a new course on Nanostructured Sensor Materials and Devices to the academic curriculum of her department a few years ago so as to introduce materials science and engineering students to this exciting world of nanostructures and miniaturized machines. This endeavor became a learning experience for her too, which led to the book that you are holding in your hands (or browsing in your electronic reader).

Acknowledgments

This book resulted from the collective effort of my research group at the Center for Nanomaterials and Sensor Development at the State University of New York, Stony Brook over the past 5 years to produce new knowledge on the synthesis, manufacturing, characterization, and testing of novel nanomaterials for sensors and biotechnology. Funded by a NSF award for Nanoscale Interdisciplinary Research Teams, a whole new research activity on novel nanomaterials flourished, producing breakthrough results, such as the "extreme" aspect ratio single crystal semiconducting nanowires or the first handheld NO breathanalyzer for monitoring oxidative stress. There are many people responsible for the successes presented here, starting from the Program Manager at the NSF's Ceramics Program, Dr. L. Madsen, who became a mentor to the me early on, to the undergraduate and graduate students who carried out the various research projects, to the collaborators and visiting scientists joining this adventure. In particular, thanks go to Dr. A. Prasad Kapaleeswaran who first produced a *selective ammonia sensor based on nanomaterials*, Dr. M. Karadge who skillfully produced CuO *nanogrids*, Ms. K. Sawicka who won a finalist position at the National Inventors' Competition for her work on *electrospun bio-nanocomposites*. The nanocomposites work is covered in chapter 4, parts of which are reproduced with permission from the Journal of Nanoparticle Research. Mrs. S. Gadre who worked on *bio-doped (non-transparent) metal oxides*; Chapter 3 is reproduced by permission of the American Ceramic Society based on a review of the field of ceramic hybrids. Continuing on, Dr. K. (Iyer) Kalyanasundaram is acknowledged for her major contribution to the development of an *electronic olfaction system based on nanosensors*, as well as for making *selective nanoprobes for signaling metabolites*. Parts of Chapter 2, which describes resistive metabolite sensing by nanomaterias, are reproduced with permission from the publisher of "Science and Technology of Chemiresistive Gas Sensors" book. It was Dr. L. Wang who produced the first *handheld detector for selective acetone measuring in exhaled breath*. Mr. D. Han and Mr. K. Ramachandran both had significant contributions in preparing *electrospun 3D scaffolds for tissue engineering* and Dr. A. Bishop-Haynes produced *electroactive polymers for nanomedicine* applications. Mr. R. Xue assisted with the editing of this book. Collaborators Abroad: Professor S. Pratsinis and Dr. A. Teleki (ETH Zurich, Switzerland), Professor G. Sberveglieri and Dr. E. Comini (Univ. of Brescia, Italy), Professor H. Doumanidis (Univ. of Cyprus), Dr. C. Balazsi and Dr. J. Pfeifer (Hungarian Academy of Sciences), Dr. H. Haneda, and Dr. N. Ohashi (NIMS, Japan) have also supported and contributed to the work presented in this book in one way or another and are greatly appreciated. Furthermore, US-based collaborators from academia, industry and from National and Defense Laboratories, Professor S.A. Akbar (The Ohio State University), Dr. M. Frame (SUNY Stony Brook), Dr. G. Gaudette (WPI), Dr. D. Kubinski and Dr. J. Visser (Ford SRL), and Dr. H. Scheuder-Gibson (Natick Army Laboratories), have also been instrumental to the success of our research and are dearly acknowledged. In closing, sincere thanks are due to my maternal grandmother, Mrs. Pelagia Micha, for the unconditional love, support, guidance and insight she has always offered to me, to my spouse, Antonios Michailidis, for his encouragement to bring this book to completion, and especially to my beloved child, Nectarios (Aris), who brought new light and happiness to my life and new meaning to my work. Sincere thanks and God bless you all.

Permissions

The author has been granted permissions to use materials from the following publications in this book:

Several parts of Chapter 2 have been reprinted from the book *Science and Technology of Chemiresistor Gas Sensors*, edited by D. K. Aswal and S. K. Gupta. Parts of Chapter 4 have been printed from *Nanostructured Metal Oxides and Their Hybrids for Gas Sensing Applications* by K. Kalyanasundaram and P. I. Gouma, pp. 147-176, 2007, with permission from Nova Science Publishers, Inc.

Chapter 3 has been reprinted from "Biodoped ceramics: Synthesis, properties, and applications" by S. Y. Gadre and P. I. Gouma, *J. Am. Ceram. Soc.* **89**(10), pp. 2987–3002 (2006). Copyright: The American Ceramic Society, with permission from Wiley.

Parts of Chapters 4 and 5 have been reprinted from "Electrospun composite nanofibers for functional applications", by. K. M. Sawicka and P. I. Gouma, *J. Nanoparticle Research* **8**(6), pp. 769–781 (2006). With permission from Springer Science and Business Media.

Glossary

Alternative energy: Forms of energy sources other than conventional fuels, such as solar power, wind power, nuclear power, etc.

Artificial organ: Man-made device implanted into the human body intending to replace a natural organ

Band gap, for semiconductors: It is the energy difference between the bottom of the conduction band and the top of the valence band, i.e. the energy range where no electron states exist

Bio-doped material: Composites of biomolecules and other inorganic or organic materials, such as ceramics or polymers where the biological component retains/enhances it's functionality

Biosensor: An electronic device that receives a biochemical signal and converts it to a measurable output

Ceramic: Inorganic, non-metallic solid; metal oxides are the most common ceramic materials

Cell: The smallest unit of an organism

Conducting polymer: An organic material that (inherently) exhibits electrical conductivity

Data processing: The mathematical process converting a set of data into useful information

Debye length: The distance in the crystal over which significant charge separation may occur

Depletion layer: The region in a (doped-)semiconductor that is depleted from electric charge due to external effects (i.e. chemisorption of gas molecules on the material's surface)

Drug delivery vehicle: Novel materials allowing therapeutic agents to be stored and released in a controlled manner

ECM: Extracellular matrix-the nature's scaffold for tissue, bone, and organ growth

Electronic energy levels: Regions of specific energy on which electrons are distributed in an atom

Electronic Nose System: Sensor materials, device architecture, intelligent signal processing routines

Electronic olfaction system: Intellligent chemical sensor array system for odor classification

Electrochemoactuator: A device that converts a chemical stimulus (energy) to a physical output (charge transfer / change in electrical resistance)

Electrospinning: A manufacturing process producing non-woven fiber mats through the application of a strong electric field to a polymer-based solution or melt

Entrapment: The encapsulation of organic species in a porous material

Hybrid material: Organic-inorganic composite

Implant: A material (natural or synthetic) that is used to replace damaged tissue/bone

Magneli phases: Families of non-stoichiometric oxides first discovered by Magneli

Metal oxide: A ceramic material consisting by one or more metal cations and oxygen ions

Nanobelt: 2D nanostructure

Nanomanufacturing: Any process producing nanomaterials

Nanowire: 1D nanostructure

Pattern recognition: Software that categorizes data sets according to specified criteria

Polymorph: A phase of the material with distinct crystallographic configuration from another with the same chemical composition

Polymorphism: The effect of a material existing in different crystallographic arrangements

Reaction kinetics: Chemical kinetics (dynamics) of a process, such as diffusion profile, etc

Resistive gas detection: The process in which a chemical stimulus is converted to changes in the electrical conductance resistance of a gas sensitive material

Scaffold: The structural basis for cell growth, proliferation, and differentiation

Tissue engineering: The art of bone/tissue/organ regeneration

VOC: Volatile organic chemical compounds

Contents Summary

Chapter 1, the Introduction to the book, deals with the definitions of the important, keyword terms used throughout this book such as nanomaterials, sensors, biotechnology, and nanomedicine. Chapter 2 introduces resistive gas detection and the effect of nanostructures in obtaining extreme gas sensitivity and for stabilizing non-equilibrium phases offering gas specificity. Furthermore, key processing methods of chemosensing nanomaterials are discussed, and emphasis is paid to nanowire fabrication and use. Case studies of nanosensor technologies are presented. Chapter 3 introduces hybrid (organic-inorganic) nanomaterials–that is bio-doped metal oxide nanosystems for chemosensing. This is a new class of materials expecting to impact not only bio-related diagnostics but also a plethora of other fields, from environmental monitoring and remediation to alternative energy. Next, in Chapter 4, the nanomanufacturing technique of electrospinning is covered in some detail, as it is a unique processing technology that enables the formation of almost all nanomaterials used in sensing and biotechnology, from semiconductor nanowires to drug delivery vehicles and artificial scaffolds for organ growth. Chapter 5 is where nanomedicine applications of nanomaterials are presented, including electronic olfaction systems and breath analyzers, the 3D fibrous scaffold approach for tissue engineering and the polymeric electro-chemo-actuators. The book concludes with an overview and insights for the future.

Contents

Chapter One

Introduction

1.1 DEFINITIONS

Learning about nanomaterials for sensors requires that the keyword terms be defined up front. Naturally, the first important term that needs to be determined is *nanomaterial*. What is this? Most people familiar with the term will respond: the material that has at least one dimension to be between 1–100 nm in size (whereas 1 nm equals 10^{-9} m. Fair enough you might say; but this is only one part of the definition. According to the **NNI** (National Nanotechnology Initiative) the material that has at least one nano-dimension has to also exhibit size-dependent properties, to be called a nanomaterial. Aluminun nanocrystals have a melting point that drops (i.e. the solid to liquid transition temperature decreases) with decreasing grain size of the material. Similarly, TiO_2 (titanium dioxide or titania) is known to transform from anatase to rutile at 1200°C when in bulk form; however, this transition occurs at temperatures as low as 400°C for 10 nm nanocrystals of anatase. Furthermore, the nature of the later transition (a "massive"-type transformation involving oriented attachment) results in bulk microcrystals of rutile. When a nanocrystalline anatase film was first used as a CO sensor operating at 400°C it behaved as a bulk rutile sensor, because it had transformed to bulk rutile crystals. It was this unexpected result that introduced the author to the surprises of *nanoscience* and *nanotechnology*.

To stay within the theme of this section, nanoscience is the study of materials systems and phenomena involving nanomaterials; whereas nanotechnology is the engineering of nanomaterials and miniaturized devices based on them. Combining nano-science and nano-engineering with medicine, the new field of *nanomedicine* has evolved, defined as medical diagnosis, monitoring, and treatment at the level of single molecules or molecular assemblies that provide structure, control, signaling, homeostasis, and mobility in cells. It is a very important research direction in order to understand the cellular mechanisms in living cells, and to develop advanced technologies for the early diagnosis and treatment of various diseases.

This book focuses on two key applications of nanomaterials: **sensors** (bio-/chemical detectors, in particular) and **biotechnology** (regenerative

Nanomaterials for Chemical Sensors and Biotechnology by Pelagia-Irene (Perena) Gouma
Copyright © 2010 by Pan Stanford Publishing Pte Ltd
www.panstanford.com
978–981-4267-11-3

nanomedicine, and non-invasive diagnostics). Sensors are devices that receive a physical, chemical or biological input (stimulus) and which provide an output signal in response to this stimulus. They consist of an active element (detector) and a transducer. The active element in the work reported in this book is a nanomaterial, organic, inorganic, or composite/hybrid. The transducer converts the receiving information (chemical composition of gas or bio-related compound) to another measurable signal (e.g. change in electrical resistance). Thus, there are myriads of different sensor technologies for gas detection alone. The focus will remain on resistive type sensors, as they are the most promising in terms of rapid response to the presence of the chemical analyte of interest, the most versatile and economic to fabricate, and the only ones having the potential to be selective to the gas of interest in the presence of interfering compounds. The active elements for these are either ceramics, or (mostly electroactive) polymers, or a combination of both. Biotechnology, within the context of this book, addresses the need for biodetectors, bio-mimicking synthetic skin, bone, and organ implants, chemomechanical actuators, and artificial olfaction-based non-invasive diagnostic kits.

REFERENCES

[1] National Nanotechnology Initiative (www.nano-gov).
[2] American Academy of Nanomedicine (www.aananomed.org).

Chapter Two

Resistive Gas Sensing Using Nanomaterials

2.1 SIZE MATTERS AND THEREFORE 'NANO' MATTERS FOR GAS SENSING

A 'nano'meter is 10^{-9}m. Nanomaterials are defined as the materials having at least one of their dimensions ≤ 100 nm. Thus we may visualize them as structures produced by reducing one, two, or three dimensions of a bulk material, thereby resulting in 2D nanolayers, 1D nanowires or 0D nanoclusters. Such length scales are close to atomic sizes (see Fig. 2.1) and at the nanoscale, the physical, chemical, and biological properties of materials differ in fundamental and valuable ways from the properties of individual atoms and molecules or bulk matter (Lieber, 1998). A wealth of new and interesting phenomena such as size dependent emission or excitation, quantized conduction, single electron tunneling (SET), metal to insulator transition to mention a few, are associated with bulk to nano transition. Quantum confinement of electrons by potential wells of nanometer sized wells may provide one of the most effective methods of modifying the electrical, optical, thermoelectric and magnetic properties of materials.

It all started with the talk by Richard Feynman —"There is plenty of room at the bottom" at the American Physical Society meeting at California's Institute of Technology, during which he outlined the advantages of having better control over things at smaller dimensions which spurred the drive for miniaturization and beating Moore's law. The book by Eric Drexler titled "Engines of Creation-The Coming of Era of Nanotechnology" came next (in the early 1990's). With the discovery of a new form of carbon- the nanotube in 1991 by Sumio Iijima and the Nobel Prize winning discovery of C60-fullerene, it seemed that the "nanoage" had really started. This paved the way for an avalanche of research and developmental activities on not only C60 but other nanomaterials as well (see Table 2.1).

At such small length scales most of the atoms are surface atoms, thus significantly increasing the effective number of sites available for reactions. Increase in surface area to volume ratio with decrease in grain size also is very important in the purview of sensing. Thus reducing the grain size plays a very important role in applications that involve surface reactions like catalysis, chemical gas sensing etc.

Nanomaterials for Chemical Sensors and Biotechnology by Pelagia-Irene (Perena) Gouma
Copyright © 2010 by Pan Stanford Publishing Pte Ltd
www.panstanford.com
978–981–4267–11–3

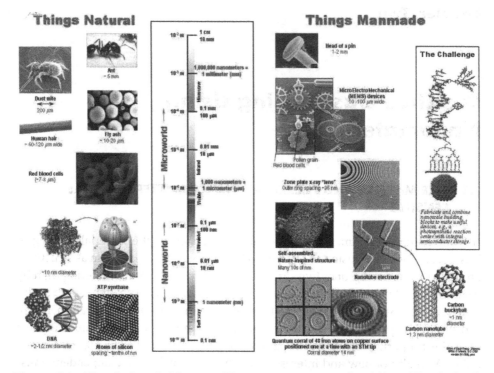

Figure 2.1. The Scale of things. [From www.nano.gov/html/facts/The_Scale_of_Things.html, National Nanotechnology Initiative- NSF.] For color reference, turn to page 143.

Table 2.1 Common Dimensionalities of Nanomaterials.

	Size (approx)	Materials
(a) Nanocrystals and clusters (quantum dots)	Diameter 1–10 nm	Metals, Semiconductors, Magnetic Materials
Other Nanoparticles	Diameter 1–100 nm	Ceramic Oxides
(b) Nanowires	Diameter 1–100 nm	Metals, Semiconductors, Oxides, Sulfides, Nitrides
Nanotubes	Diameter 1–100 nm	Carbon, Layered Metal Chalcogenides
(c) Two dimensional arrays (of nanoparticles)	Several $nm^2 - \mu m^2$ in area	Metals, Semiconductors, Magnetic Materials
Surfaces and thin films	Thickness 1–1000 nm	Several materials
(d) 3 dimensional structures (superlattices)	Several nm in all three dimensions	Metals, Semiconductors, Magnetic materials

2.2 NANOSTRUCTURED METAL OXIDES FOR CHEMIRESISTORS

Metal oxides have been used nearly for four decades for gas sensing applications. The basic principle behind the gas sensing mechanism by metal oxides is the change in their electrical resistance on exposure to a gas, due to electronic exchange. The discoveries by Seiyama *et al.* in 1962 that ZnO thin films exhibit changes in their electrical conductivity with small amounts of reducing gases and the same year by Taguchi *et al.* that SnO_2 partially sintered pellets respond similarly were the beginning for what has been a rapid gas sensor developmental phase. Since 1968 Taguchi sensors have been mass-produced and with the establishment of Figaro Engineering Inc. in 1969 the SnO_2 sensors have been commercially available. What started with thick films and pressed pellet bulk sensors has now evolved in to novel nano-architectures for gas sensing applications that have sensitivities down to ppb levels.

Conducting polymers also offer the feasibility of their use for resistive gas sensing, as discussed below, metal oxides are mainly attractive, because of their ease of fabrication, advantage of easy integration with circuits and MEMS devices, better structural and chemical control, responses down to ppb levels of gases and relative inexpensiveness.

2.2.1 Nanostructured Metal Oxides

Having looked at size dependence of properties, it is important to relate them to how they affect the gas sensing behavior of metal oxides that have nano grain size. The following points are a few of the "size dependent" properties that affect the electronic properties of the metal oxides.

Figure 2.2 shows the grain size dependence of sensitivity for SnO_2 films exposed to CO and H_2 (Yamazoe's group, 1991). Similarly Fig. 2.3 shows the response of an In_2O_3 film to O_3 (ozone) as a function of grain size of the In_2O_3 film. Sensitivity is calculated as the ratio of resistance in gas to the resistance in air in the case of oxidizing gases and in the case of reducing gases it is calculated as the ratio of resistance in air to resistance. A two to three order increase in sensitivity was observed in the case of the In_2O_3 films when the grain size was decreased from 60–80 nm to 10–15 nm (Korotchenkov *et al.*, 2004).

The other factor that becomes predominant at smaller grain sizes is the depletion layer depth or called otherwise the Debye Length (L or 'λ'). It is a measure of field penetration in to the bulk (the mechanism of formation of this layer will be provided in the next section). For most nanostructures, the value is comparable to their diameter (in the case of spherical particles/nanotubes/nanowires), or their width (in the case of nanobelts and other flat nanostructures). Under such conditions the surface chemical processes strongly influence the electronic properties. A nanowire can go from a completely insulating state to a completely conducting state (Kolmakov *et al.*, 2004).

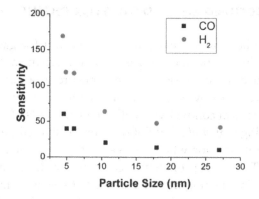

Figure 2.2. The effect of particle size on the gas sensitivity for a SnO$_2$ sensor exposed to CO and H$_2$. [Reprinted from Sensors and Actuators B, 3(2) pp. 147–155, C. Xu, J. Tamaki, N. Miura, and N. Yamazoe, Grain size effects on gas sensitivity of porous SnO2-based elements © (1991), with permission from Elsevier.] For color reference, turn to page 143.

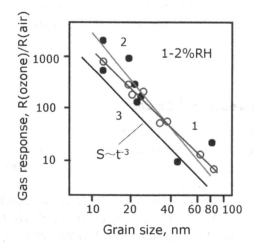

Figure 2.3. Influence of In$_2$O$_3$ film grain size on the gas response to ozone. [Reprinted from Thin Solid Films, 460, G. Korotcenkov, V. Brinzari, A. Cerneavschi, M. Ivanov, V. Golvanov, A. Cornet, J. Morante, A. Cabot and J. Arbiol, The influence of film structure on In$_2$O$_3$ gas response, p. 315, © (2004), with permission from Elsevier.] For color reference, turn to page 144.

Quantum confinement effects become pronounced when dimensional anisotropy sets in.

In order to understand this better, it is necessary to obtain a background on the gas sensing mechanism of metal oxides. Metal oxides in the simplest way can be described as having metal atoms (may be one similar or dissimilar metal) bound to oxygen atoms. It is the surfaces of these metal oxides that are most important of chemical gas sensing.

2.2.2 Basic Mechanisms of Gas Sensing Using Semiconductors

The most quoted model to explain the resistance change in a metal oxide semiconductor sensor is that, in air, oxygen adsorbs on the surface, dissociates to form O^-, where the electron on the oxygen, is extracted from the semiconductor. This electron extraction tends to increase the resistance (assuming an n-type semiconductor (whose majority charge carriers are electrons)). In the presence of a combustible gas, like say H_2, the hydrogen reacts with the adsorbed O^-, to form water and the electron is re-injected in to the semiconductor, tending to decrease the resistance. A competition results between the oxygen removing the electrons and the combustible gas restoring these electrons. So, the steady state value of resistance of the metal oxide depends on the concentration of the combustible gas. This could be illustrated in the following way, by considering the competing reactions:

$$O_2 + 2e^- \longrightarrow 2O^- \tag{2.1}$$

$$H_2 + O \longrightarrow H_2O + e^- \tag{2.2}$$

The more the H_2 present the lower the density of O^-, the higher the electron density in the semiconductor, and thus lower the resistance.

Another model that may exist or co-exist is that the combustible gas, if chemically active, extracts a lattice-oxygen from the metal oxide, leaving vacancies that act as donors. The oxygen from the air tends to re-oxidize the metal oxide, removing the donor vacancies. Thus, there is a competition between the oxygen removing the vacancies and the combustible gas producing donor vacancies. The density of donor vacancies (and therefore the resistance) depends only on the concentration of combustible gas because the oxygen partial pressure is constant (as when operating in air) (Sze 1994).

2.2.3 Surface States in Ionic Crystals like SnO$_2$

In an ionic crystal like SnO_2, both anions and cations have poor coordination at the surface. The positively charged Sn ions at the surface have an incomplete shell of negative oxide ions around them. With too few negative ion neighbors, the positively charged ions are more attractive to electrons. So their conduction-band-like orbitals can be at a lower energy than the conduction band edge and can 'capture' electrons from the bulk. They also can bond well with a 'basic' molecule such as OH^-, which has an electron pair to give to the bond. The surface anions on the other hand do not have their quota of positive ions around them so their anionic-like orbitals can be at an energy level higher than the valence band edge. They can capture holes or give up electrons to the bulk. They can also bond well with an acid molecule like H^+ that has a pair of unoccupied orbitals. Actually at low temperatures it is expected that the ionic solids will normally be covered by chemisorbed water, the OH^- bonded to the surface at cationic and the H^+ at anionic sites. At higher temperatures these water molecules can be driven off leaving the active sites open for interaction with bases or acid gaseous species or with electron-donor or electron-acceptor gaseous molecules.

The electronic energy levels in the bandgap are termed "surface states" and for semiconductors there are both donor and acceptor levels present on the surface. If there no carrier exchange between the surface states and the bulk of the semiconductor the bands remain flat and the energy band model is as shown in Fig. 2.4. Figure 2.5 illustrates what happens when the electrons move from the region of high EF, the near surface region of the semiconductor, to a region of lower EF, the surface states. This separation leads to a double layer voltage that either raises or lowers the energy of the surface states. This movement of bands near the surface is called "band bending".

The electrical double layer formed may be of three types depending on the movement of electrons in to and out of the double layer. They are:

(1) Accumulation layer-This type of layer forms when electrons are injected in to an "n-type" semiconductor (Fig. 2.6)
(2) Depletion or Exhaustion or space charge layer-This forms when electrons are extracted from the conduction band of the n-type semiconductor (Fig. 2.7)
(3) Inversion layer-This type of layer forms in FET devices when a very strong oxidizing agent such as fluorine is present.

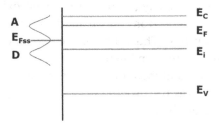

Figure 2.4. 'Flat Band' condition-No charge exchange between the surface states and the bulk. [Figures 2.4–2.7 Reprinted from Semiconductor Sensors, S. M. Sze, 1st Edition, 1994 (John Wiley & Sons Inc.), with permission from John Wiley & Sons Inc.] For color reference, turn to page 144.

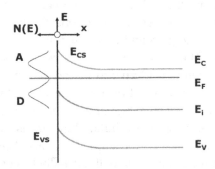

Figure 2.5. 'Band Bending'-where the electrons from the semiconductor surface have moved to the surface states[3]. For color reference, turn to page 144.

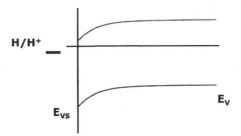

Figure 2.6. Formation of an accumulation layer between the electropositive surface species and the negatively charged semiconductor. For color reference, turn to page 145.

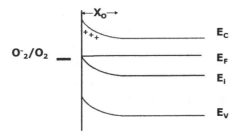

Figure 2.7. Formation of a depletion layer between the negatively charged surface species and the positive donor ions. For color reference, turn to page 145.

Of these, the most important type of layer for gas sensing is the depletion layer. Hence it will be discussed in a little more detail. It was said earlier that the depletion layer forms in a 'n-type' semiconductor when electrons are extracted from it. Similarly in a p-type semiconductor this type of layer will form when holes are extracted from it. In an n-type semiconductor the double layer forms between the negatively charged surface states and the positively charged donor (immobile) ions in the bulk of the semiconductor. In a p-type semiconductor the double layer forms between the positively charged surface states and the negatively charged acceptor ions in the bulk.

2.2.4 *n-p* Type and *p-n* Type Transitions in Semiconductor Gas Sensors

Semiconductor type transitions can occur during sensing. The exact mechanisms of transition are not exactly known though there are two possible explanations for it. The first one is an inversion layer formation on the surface that locally causes a transition from an *n-p* type or *p-n* type transition depending on the adsorbate. An inversion layer forms on the surface of an *n*-type semiconductor in the presence of a strong oxidizing agent resulting in the formation of an acceptor surface state. If the surface state energy level is close to the valence band edge then to bring the Fermi level close to the surface state, the surface Fermi energy must be close to the valence band. In such a situation the acceptor surface state is so low in the band diagram that it extracts electrons from the valence band leaving a substantial hole concentration. This results in a local n-p type transition.

In the other case, the common mechanism that has been suggested is the formation of oxygen vacancies due to a loss of local stoichiometry (Gregory *et al.*, 2002). In general, the conductivity of a semiconductor is given as,

$$\sigma = -qn\mu_n + qp\mu_p \qquad (2.3)$$

where n and p are the electron and hole concentrations respectively, q is the associated charge and μ is the associated charge mobility. When the concentration of either of the charge carriers becomes larger than the other, there is a shift in the type from p to n or n-to p. It has been found that the values of n and p depend on the generation of inter-band traps due to the formation of vacancies or impurity substitution. It has been found in MoO_3 that there is a p-n type transition (Prasad *et al.*, 2003). This might be due to the formation of oxygen vacancies that leads to excess electrons or incorporation of oxygen atoms in to these vacancies that leads to excess holes. When either of these values exceeds a threshold level there is a transition from one type of conduction to other.

2.2.5 Importance of 'nano'

Calculations for the results below have been originally carried out in (Barsan, 1994) and have been derived from (Barsan *et al.*, 2001). For grains/crystallites large enough to have a bulk region unaffected by the surface phenomena i.e. when the grain diameter d $\gg$ Debye length λ_D the surface charge carrier density n_s, is given by

$$n_s = n_b\exp(-qV_s/k_BT) \qquad (2.4)$$

For the limiting case when the crystallite size d is $\leq \lambda_D$, the activation energy related to the Debye length as

$$\Delta E \sim k_B.T.\{R/2\lambda_D) \qquad (2.5)$$

where, R is the radius of the cylindrical filament produced by sintering small grains. If the value of ΔE is comparable to thermal activation, then we have a homogeneous electron distribution in the filament and flat band conditions.

Some of these parameters like the concentration of free charge carriers (electrons), the Hall mobility μ, the Debye length λ_D, and the mean free path of the free charge carriers λ have been calculated for single crystal SnO_2 surfaces for various temperatures (Barsan *et al.*, 2001). (n_b is the concentration of free charge carriers (electrons), μ_b is their Hall mobility, λ_D is the Debye length, and λ is the mean free path of free charge carriers).

If ΔE is comparable to the thermal energy then a homogeneous electron concentration is attained in the grain and leads to the flat band case. For grain sizes lower than 50 nm, it has been shown that complete depletion of charge carriers occurs inside the grain and a flat band condition results for almost all temperatures except a few.

Table 2.2 Bulk and surface parameters of influence for SnO_2 single crystals (adapted from Barsan *et al.*, 2001).

T (K)	400	500	600	700
n_b	1	11	58	260
μ_b $(10^4$ $m^2/$ $(V_s))$	178	87	49	31
λ_D (nm)	129	43	21	11
Λ (nm)	1.96	1.07	0.66	0.45
$\Delta E/(k_B T)\mid_{(R=50nm)}$	0.34	0.77	1.08	1.49

2.3 SELECTIVITY IN GAS-OXIDE INTERACTIONS — EFFECT OF OXIDE POLYMORPHISM

In the author's research it has been surprisingly discovered that the ability of selective detection of a specific gaseous analyte in the presence of interfering gas mixtures, i.e., sensor selectivity, may be determined by the careful selection of the crystalline polymorph, i.e. specific crystallographic phase, of a stoichiometrically pure metal oxide to be used for sensing. This is due, in large part, to the fact that semiconducting metal oxides show surface sensitivity to re-dox reactions involving gases. The local environment of active sites of metal oxides used for gas adsorption, hydrogen extraction, or oxygen addition, and the orientation of the surface containing these active sites, differ for different phases of a given metal oxide, and thus the catalytic behavior of these phases is different. For example, oxide phases may expose different types of oxygen vacancies at their surfaces; surface oxygen vacancies formed under reduction conditions have an influence on gas adsorption. These vacancies may result in slightly reduced metal oxide surfaces which undergo re-oxidation by gaseous oxygen, which is adsorption based sensing, or they may order and give rise to crystallographic shear structures that accommodate non-stoichiometric metal oxide compositions known as Magneli phases which is reaction based sensing. The presence of ordered vacancies and crystallographic shear structures thus provides a mechanism for selective oxidation and may be utilized in selecting appropriate metal oxides.

Metal oxides may be classified into three groups according to their crystallographic characteristics (Gouma, 2003). The gases detected by these metal oxides are oxidizing or reducing gases and similarly fall into three categories. The first group of metal oxides, classified here as "rutile structured" metal oxides, possess a rutile crystallographic configuration similar to that found in TiO_2. The rutile structure is tetragonal, but in some cases it has been described as a distorted hexagonal close packed oxide array with half the octahedral sites occupied by the metal. Alternate rows of octahedral sites are full and empty. The rutile structure is regarded as an ionic structure. Examples of rutile structured metal oxides include TiO_2, SnO_2, CrO_2, IrO_2, β-MnO_2, etc.

The second group of metal oxides, classified herein as "rhenium oxide structured" or "ReO$_3$-type" metal oxides, possess a cubic structure akin to that found for rhenium oxide (ReO$_3$), which is closely related to the structure found in perovskite (CaTiO$_3$). The unit cell of the crystal contains metal atoms at the corners with oxygen at the center edges. Metal oxides which form this structure include WO$_3$, β-MoO$_3$.

The third group of metal oxides, classified herein as "α-MoO$_3$-type" metal oxides have a unique, weakly bonded 2D layered structure; α-MoO$_3$ and hexagonal WO$_3$ are typical representatives of this group.

Gases that may be detected by the metal oxide sensing elements may similarly be placed into three categories (Gouma, 2003). Type I gases are nitrogen-lacking reducing gases including, but not limited to, CO, alcohols, and hydrocarbons. Type II gases are nitrogen-containing reducing gases including, but not limited to, NH$_3$ and amines. Finally, Type III gases are oxidizing gases including, but not limited to, O$_2$, NO, NO$_2$ etc.

Rutile structured metal oxides are selective in their sensitivity to the nitrogen-lacking reducing gases (Type I); the ReO$_3$-type metal oxides are selective in their sensitivity to the oxidizing gases (Type III); and the α-MoO$_3$-type metal oxides are selective in their sensitivity to the nitrogen-containing reducing gases (Type II).

The first step in preparing a sensor with selectivity for a specific gas is determining the reducing or oxidizing nature of the gas being tested, and then selecting a metal oxide for use in a sensor, on the basis of the crystal structure of the metal oxide (see Fig. 2.8). This selection is based on analyzing the primary nature of the gas-metal oxide interactions, i.e., chemisorption vs. reaction-based sensing. The next step is to match a specific oxide within a designated group to a specific gas of a certain type. After initially selecting the metal oxide for the

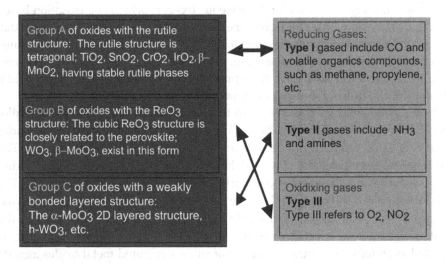

Figure 2.8. Semi-empirical map of Gas-Oxide Interactions. [From Gouma, *Rev. Adv. Mater. Sci.*, 5, pp. 123–138, 2003.] For color reference, turn to page 145.

specific gas, the appropriateness of the specific metal oxide crystal in a sensor for the specific gas may be easily determined utilizing analytical techniques that provide chemisorption related information (e.g. Raman spectroscopy conducted on FTIR optical instrument). The electronic structure of the gas to be detected also needs to be considered, especially with respect to the ease of exchanging electrons with the metal oxide surface. Crystal growth and patterning techniques allow for growing metal oxide along a preferred crystallographic orientation, thus optimizing the configuration of the metal oxide crystals in the sensor.

There are several other factors that may influence the selection of a given metal oxide for selective gas sensing, and these include the (thermal) stability of the sensor at the operating temperature, the structural stability of the chosen metal oxide phase, and the temperature dependence of the sensing process. These variables may be easily accounted for once the proper choice of metal oxide phase has been made. In addition, multisensor arrays of inherent specificity to different gases may be constructed.

2.4 NANOSTRUCTURED METAL OXIDE FABRICATION

This section will try to provide a brief outline of the methods commonly used for the fabrication of nanostructured metal oxides. There are conventional methods for fabrication of nanostructured semiconductor sensors- thick and thin film production methods. Some unconventional metal oxide nanostructures like nanobelts, nanowires, nanodiskettes will also be discussed along with the general approaches for their synthesis.

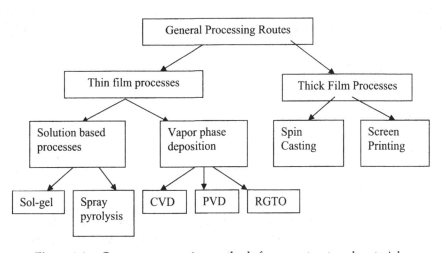

Figure 2.9. Common processing methods for nanostructured materials.

2.4.1 Conventional methods

The conventional methods are outlined in the schematic below. A brief overview of each of these techniques would also be provided.

2.4.1.1 Thin Film Technologies

I. Vapor Deposition-Nanostructure Synthesis Using PVD and CVD   A variety of oxides such as ZnO, Ga_2O_3, In_2O_3, CdO, PbO_2 have been fabricated using a solid-vapor process. It involves vaporizing a powder source material at an elevated temperature, and the resultant vapor condenses to form the desired product under specific operating conditions (temperature, pressure, atmosphere, substrate etc.).

Chemical Vapor Deposition (CVD) involves exposing a substrate of choice to a mixture of volatile precursors that react and/or decompose on the substrate to give the desired product. A wide variety of CVD techniques that are in use are as follows (Park *et al.*, 2003):

(1) Atmospheric pressure CVD (APCVD)
(2) Atomic layer CVD (ALCVD)
(3) Low pressure CVD (LPCVD)
(4) Metal organic CVD (MOCVD)
(5) Microwave plasma assisted CVD (MPCVD)
(6) Plasma enhanced CVD (PECVD)
(7) Metal organic MBE (ultra high vacuum MOCVD)

Physical vapor deposition (PVD) uses physical means rather as opposed to chemical vapor deposition techniques. The various techniques are:

(1) Sputtering
(2) Evaporation
(3) Pulsed LASER deposition

Both PVD and CVD offer enormous amount of control over the film thickness, stoichiometry and microstructure. Hence these films usually have highly controlled properties. But the only disadvantage is that these film deposition techniques are slightly expensive.

Specifically, a vacuum deposition method such as evaporation, plasma assisted chemical vapor deposition, or a sputtering method may be used for forming the thin film metal oxide sensors. In the sputtering method, predominantly neutral atomic or molecular species are ejected from a target, which may be formed from the material to be deposited, under the bombardment of inert gas positive ions, e.g., argon ions. The high-energy species ejected will travel considerable distances to be deposited on the substrate held in a medium vacuum, e.g. 10^{-4} to 10^{-2} mbar. The positive ions required for bombardment may be generated in a glow discharge where the sputtering target serves as the cathode electrode to the glow discharge system. The negative potential (with respect to ground and the glow discharge) is maintained in the case of insulating target materials by

Table 2.3 Common techniques of CVD and PVD thin film processing.

CVD	PVD	
	Sputtering	Evaporation
• Thermal CVD • Plasma enhanced CVD • Laser induced CVD • Electroless plating • Spray Pyrolysis • Melt dipping • Liquid quenching • Deposition of organic polymers and emulsions	• Reactive sputtering • Cathode sputtering • IBD • Ionized cluster beam • Plasma decomposition	• MBE • Thermal evaporation • Reactive evaporation • Ion plating • Arc evaporation • LASER evaporation

the use of radio frequency power applied to the cathode, which maintains the target surface at a negative potential throughout the process. DC power may be applied when the target is an electrically conducting material. The advantage of such techniques is that control of the target material is greatly enhanced, and the energy of the species ejected is very much higher than with evaporation methods e.g. typically 1 to 10 eV for sputtering as compared with 0.1 to 0.5 eV for evaporation methods.

In magnetron sputtering processes, the plasma is concentrated immediately in front of the cathode (target) by means of a magnetic field. The effect of the magnetic field on the gas discharge is dramatic. In that area of discharge where permanent magnets, usually installed behind the cathode, create a sufficiently strong magnetic field vertically to the electric field, secondary electrons resulting from the sputter bombardment process will be deflected by means of the Lorenz force into circular or helical paths. There is an increase in plasma density and a considerable increase in deposition rate.

Bias sputtering (or sputter ion plating) may be employed as a variation of this technique. In this case, the substrate is held at a negative potential relative to the chamber and plasma. The bombardment of the substrate by Argon ions results in highly cleaned surfaces. Sputtering of the target material onto the substrate throughout this process results in a simultaneous deposition/cleaning mechanism. This has the advantage that the interfacial bonding is considerably improved. In sputter ion-plating systems the substrate is held at a negative potential. In this case, the relative potentials are balanced to promote preferential sputtering of the target material. The target voltage will be typically less than 1 kV, dependant on system design and target material. The substrate may be immersed in its own localized plasma dependent upon its bias potential, which will be lower than that of the target. The exact voltage/power relationship achieved at either target or substrate depends upon many variables and will differ in detail from system to system.

Several heating methods exist, e.g., resistive, inductive, electron beam impingement etc., although the commonly preferred method is an ion beam source where a beam of ions impinge onto the coating material contained in a water-cooled crucible. The use of multi-pot crucibles or twin source guns, enable multiple layers and graded stoichiometry layers to be deposited with the aid of electronic monitoring and control equipment.

In ion plating, negative bias applied to the substrate in an inert gas promotes simultaneous cleaning/deposition mechanisms for optimizing adhesion, as described in the sputtering process. Bias levels of -2 kV are typically used but these can be reduced to suit substrates. As operating pressures are higher in the ion plating technique, e.g. 10^{-3} to 10^{-2} mbar, gas scattering results in a more even coating distribution. To protect the filament the electron beam gun in the ion plating technique is differentially pumped to maintain vacuum higher than 10^{-4} mbar.

In the plasma assisted chemical vapor deposition (PACVD) method, the substrate to be coated is immersed in a low pressure (0.1 to 10 Torr) plasma of the appropriate gases/volatile compounds. This pressure may be maintained by balancing the total gas flow-rate against the throughput of the pumping system. The plasma may be electrically activated and sustained by coupling the energy from a power generator through a matching network into the gas medium. Thin films have been successfully deposited from direct current and higher frequency plasmas well into the microwave range. At high frequencies, the energy may be inductively coupled depending on chamber design and electrode configuration. Typically a 13.56 MHz radio-frequency generator would be used having a rating that would allow a power density of between about 0.1 W/cm^2. Typical deposition rates for this technique can be favorably compared with those obtained by sputtering. The deposition of the thin film metal oxide may be achieved by immersing a substrate in plasma containing a metal compound, such as molybdenum or tungsten, and oxygen under appropriate processing conditions.

In addition, suitable thin film metal oxides may be formed by a sol-gel deposition method, a plasma ashing method, or a solution coating method.

II. Sol-gel method Sol-gel method (Fig. 2.10) has been used for a long time for the production of nanomaterials (Brinker and Scherrer, 1984). This is a room or slightly elevated temperature process. The process involves the hydrolysis of a metal organic compound such as a metal alkoxide (usually, or can be hexachlorides as well), or inorganic salts such as chlorides (Shieh *et al.*, 2002) to produce a colloidal sol. The hydrolysis can take place with the help of alcohol, acid or base. The sol is then allowed to age and settle. This is referred to as the gelation step.

The inorganic oxide gel is converted to an inorganic oxide glass by a low temperature heat treatment. The metal alkoxides, e.g., alkoxides of molybdenum or tungsten such as molybdenum isopropoxide or tungsten isopropoxide, may be

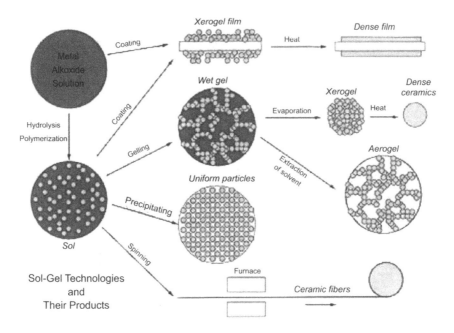

Figure 2.10. Schematic of the sol-gel processing technique (www.chemat.com).

used in combination with an alcohol, such as n-butanol, to form the metal trioxide, i.e., molybdenum trioxide or tungsten trioxide, respectively. Because these isopropoxides are reactive to atmosphere, they may be mixed in an enclosed container under a nitrogen atmosphere. The resulting sol is allowed to age and settle, and the sol may then be deposited on a substrate.

The gel coated substrate then undergoes suitable drying and firing stages to convert the coating into an inorganic oxide glass. The precise conditions with respect to temperature and residence time in the various stages of conversion are dependent upon the gel composition and its tolerance to relatively rapid changes in its environment. Porosity and integrity of the coating can be significantly affected by these stages. A suitable conversion process would include drawing the gel-coated substrate through drying ovens in which the temperature is controlled.

The versatility of the process lies the in the flexibility available for the form of the end product such as

- The sol can be coated on the substrate by either spin/dip coating to form a 'xerogel' film
- The solvent from the sol can be evaporated to precipitate particles of uniform size and then these can be screen printed
- The sol can be allowed to gel completely to obtain either a xerogel or an aerogel.
- The sol can be spun cast to form ceramic fibers

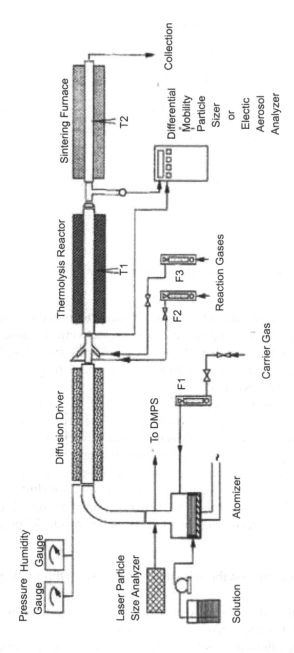

Figure 2.11. Schematic of a Spray Pyrolysis set up. [Reprinted from *Journal of the American Ceramic Society*, 76, G.L. Messing, S-C. Zhang and G. V. Jayanthi, Ceramic powder synthesis by spray pyrolysis, p.2707, with permission from Blackwell Publishing.]

III. Spray Pyrolysis This process, as shown schematically in Fig. 2.11, involves the atomization of a liquid precursor through a series of reactors, where the aerosol droplets undergo evaporation, solute condensation within the droplet, drying, thermolysis of the precipitate particle at higher temperature to form a microporous particle which then gets sintered to give a dense particle (Messing *et al.*, 1993). The advantages of using a spray pyrolysis are as follows:

- The process makes use of the wide variety of available solution chemistries compartmentalizing the solution in to unique droplets, thereby retaining very good stoichiometry on the particle surface. This is particularly useful for the synthesis of single and mixed metal oxides
- A variety of particle morphologies can be obtained such as core-shell morphologies, porous particles for catalyst support, fibers, nanocomposites, quantum dots and hollow nanoparticles, to mention a few.

There are a variety of spray pyrolysis processes, and a few of them are aerosol thermolysis, flame spray pyrolysis, aerosol decomposition, spray roasting, and aerosol decomposition.

2.4.1.2 Thick Film Technologies

Screen printing involves mixing of the high purity oxide powder with organic deflocculant additives like terpineol or propadyol along with inorganic binders like ethyl silicate/ethyl cellulose (Park *et al.*, 2003; Martinelli *et al.*, 1999). The paste should have the correct rheological properties. In addition, adherence to the substrate and correct thermal shrinkage properties are necessary to obtain a good film. The advantages of using screen printing are that it is a very efficient way to produce cheap and robust chemical sensors. Also it offers excellent control over the obtainable thicknesses.

2.4.2 Unconventional Nanostructures

Recently there have been a lot of reports of synthesis of 2D and 1D nanostructures such as nanowires, nanobelts, nanorods, nanodiskettes of metal oxides, with a large surface area to volume ratio, see Fig. 2.12. These novel nano-assemblies are expected to possess unique properties such as very high sensitivities to single molecules or few ppbs of gases, quantum confinement. Xia *et al.*, 2003, outlines the strategies for achieving one dimensional growth. It has been generally accepted that there must exist a reversible pathway between the building blocks on the solid surface and the fluid phase (liquid, vapor, gas). The building blocks also need to be supplied at a controlled rate in order to achieve uniform composition and morphology. The general strategies for achieving 1D growth can be summarized as follows:

- Use of intrinsically anisotropic crystal structure
- Introduction of a solid-liquid interface to reduce the symmetry of the seed

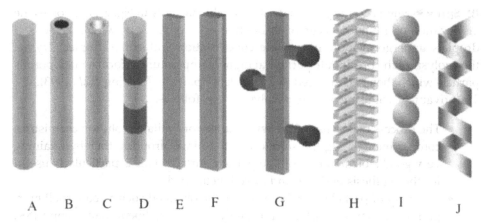

Figure 2.12. A-Nanowires and nanorods, B-Core-shell structure with metallic inner core, C-Nanotubes, nanopipes and hollow nanorods, D-heterostructures, E-nanobelts/ nanoribbons, F-nanotapes, G-nanodendrites, H-hierarchical nanostructures, I-nanosphere assembly, J- nanosprings. [Reprinted, with permission, from the Annual Review of Materials Research, Volume 34 ©2004 by Annual Reviews www.annualreviews.org.]

- Use of templates to achieve directional growth
- Use of supersaturation control to modify the habit of the seed
- Use of capping agents to control the growth of various crystal facets
- Self-assembly of 0D nanostructures
- Size reduction of 1D nanostructures

2.5 CASE STUDIES OF NANOSTRUCTURED SEMICONDUCTOR METAL OXIDES

2.5.1 3D Nanostructures

Nanostructured TiO_2 for VOC detection: Large and easily accessible surface area, high crystallinity and the ability to include noble metal doping are all requirements for TiO_2 sensor material synthesis routes. Nano- and micrometer TiO_2 particles for gas-sensing have been produced by sol-gel (Montesperelli et al.[3] 1995), oxidation of metallic titanium foil (Gouma *et al.*, 2000), laser pyrolysis (Carotta *et al.*, 1999), magnetron sputtering (Harris, 1980), supersonic cluster beam deposition (Mazza *et al.*, 2005) and ball milling of commercial powders (Savage *et al.*, 2001). An alternative and highly attractive synthesis route for nanostructured TiO_2 is flame technology which is used largely for manufacture of about 2 million tonnes/year pigmentary titania (Braun, 1997). The size, crystallinity, and morphology of flame-made TiO_2 can be controlled by changing the high temperature residence time of the particles in the flame (Pratsinis *et al.*, 1996; Wegner *et al.*, 2003). For example, silica stabilizes the anatase phase, while alumina or tin oxide promote rutile formation (Vemury *et al.*, 1995; Teleki *et al.*, 2005). Addition of low-volatility dopant precursors (e.g. for platinum) can be facilitated by flame spray pyrolysis (Strobel *et al.*, 2003) that have resulted in highly sensitive Pt/SnO_2 sensors (Madler *et al.*, 2006).

The plot with the sensing data illustrated in Fig. 2.13 compares the sensing response of flame-sprayed titania nanoparticles to two hydrocarbons, acetone and isoprene. The response curves of acetone and isoprene coincide, only at 7.5 ppm acetone gives a higher signal than isoprene (Teleki *et al.*, 2006). Hydroxyl groups are present on the surface of flame-made TiO_2 nanoparticles (Mueller *et al.*, 2003). Obee and Brown (1995) suggested that 1,3-butadiene adsorbs on TiO_2 by formation of a OH ... -electron type complex with the hydroxyl groups present on the surface. Hydroxyl groups will also interact with acetone molecules through hydrogen bonding (Coronado *et al.*, 2003). The interaction of both acetone and isoprene with hydroxyl groups on the surface of TiO_2 might explain the sensor signal similarity of the two gases.

The response times are within 2–3s for acetone and isoprene at all concentrations. The recovery time for acetone increases nearly linearly from 144 s at 1 ppm to 302 s at 7.5 ppm. Isoprene shows the same dependence of recovery time with concentration but the sensor recovers faster than for acetone. The slower recovery after acetone exposure might stem from molecular adsorption of acetone on the surface. Acetone adsorbs on TiO_2 surfaces mainly by coordination to Ti^{4+} sites via the lone electron pair on the oxygen atom of acetone forming eta1- acetone (Henderson, 2004). A thermal reaction might also occur between acetone and adsorbed oxygen species forming an acetone–oxygen complex followed by decomposition of the molecule (Henderson, 2005).

In the same study (Teleki *et al.*, 2006), it was shown that a different sensing mechanism occurs for ethanol. Upon adsorption of ethanol, the O–H bond dissociates heterolytically to yield an ethoxide and a proton (Idriss and Seebauer, 2000). As this is a reversible reaction, through protonation of the ethoxide, ethanol can be released to the gas phase again. Alternatively, the reaction on the surface can proceed, eventually forming an acetaldehyde, this reaction, however, is slow compared to the adsorption step (Idriss and Seebauer, 2000). An increase in ethanol concentration might shift this reaction equilibrium towards the formation of acetaldehyde and also increase the reaction rate, which could explain the decrease in recovery time with increasing ethanol concentration observed in our studies. Also Ferroni *et al.*, (Ferroni *et al.*, 2002) suggested that the sensing behavior of titania not only relies on interaction of ethanol with adsorbed oxygen species, but rather on the direct adsorption at semiconductor surface sites.

2.5.2 1D Nanostructures

2.5.2.1 Basic Gas Sensing Mechanism in 1D Nanostructures

Before addressing the individual metal oxides and their gas sensing properties, the gas sensing mechanism in one-dimensional nanostructures will be examined. As discussed above, the sensing mechanism of metal oxide semiconductors is one that involves the alteration of band structure by the adsorption of chemical species at

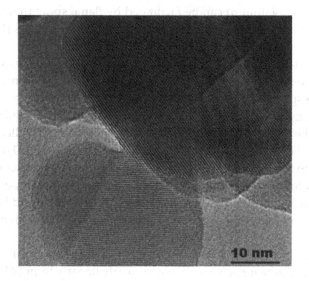

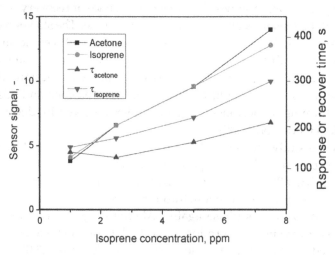

Figure 2.13. **(top)** HRTEM image of nanocrystalline flame-sprayed titania; **(bottom)** comparative plot of sensing data for isoprene and ethanol. For color reference, turn to page 146.

the grain boundaries due to redox reactions. The idea behind using one dimensional nanostructures for sensing is to eliminate the effects of bulk and have the depletion layer extend throughout the thickness of the nanostructure i.e., making the Debye length (λ_D) comparable to the grain diameter. It is a generally accepted theory that the nanostructures are small enough for the adsorption of redox species to alter the bulk electronic structure of the entire nanowire or nanobelt and not merely its surface. Summing up, it can be said that the nanostructure acts as a conductance switch, the bulk conductivity of which fully depends on its surface chemistry.

To understand this better, let us consider the following scheme. After annealing at high temperatures, the surface of the metal oxide loses its lattice oxygen (O^{2-}) and ionosorbed oxygen (O^-), leading to the formation of oxygen vacancies (Vo). The vacancies act as donors (donate electrons to the semiconductor) as shown by Eqs. (2.6–2.9), and thereby making the metal oxide an n-type semiconductor (Kolmakov *et al.*, 2003).

$$O^{2-} \longrightarrow \frac{1}{2} O_{2(g)} + V_o \tag{2.6}$$

$$V_o \longrightarrow V_o^+ + e^- \tag{2.7}$$

$$V_o^+ \longrightarrow V_o^{++} + e^- \tag{2.8}$$

$$O^- \longrightarrow \frac{1}{2} O_2 + e^- \tag{2.9}$$

Exposure to oxygen, see Fig. 2.14, recreates the surface oxygen acceptor states (as discussed above), leading to an increase in the resistance of the semiconductor. When reducing gases such as carbon monoxide, for example, react with these ionosorbed oxygen species, and releasing the trapped electron back in to the conduction band, thereby increasing the conductance of the metal oxide.

The mechanism discussed above is the same as that discussed in section 2.2, except that in the case of one-dimensional nanostructures, the thickness of the layer in which electron exchange between the bulk and surface states takes place (the depletion layer, λ_D) is comparable to the dimensions of the nanostructure itself.

2.5.3 Use of Catalysts

One of the most popular methods to improve the selectivity and sensitivity of the sensor is by the use of catalysts. There are two possible methods by which the catalysts may improve the sensitivity by modifying the intergranular contact resistance, namely

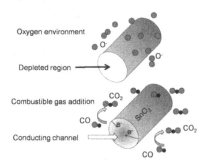

Figure 2.14. Sensing mechanism in a SnO_2 nanowire depicting the complete depletion of charge carriers inside the nanowire. [Reprinted from Advanced materials, 15, A. Kolmakov, Y. Zhang, G. Cheng and M. Moskovits, Detection of CO and O_2 using tin oxide nanowire sensors, p. 997, © 2003, with permission from Wiley-VCH Verlag GmbH & Co KG.]

- Chemical Sensitization — The catalyst particles on the surface of the support aid in the dissociation of the reducing/oxidizing gas thereby resulting in the "spillover" of the reaction products onto the surface of the support (The term support in catalytic literature refers to the ionic solid on which the catalyst is deposited. In our case it refers to the semiconductor sensor). The spilled over products react in the usual way (Morrison, 1987).
- Electronic sensitization — The catalyst is involved in electronic exchange with the gas phase and in turn with the support surface, thereby modifying the Fermi energy and the barrier height of the support surface (Morrison, 1987).

In both cases a very good dispersion of the catalyst is necessary, in order to obtain improved sensitivity and selectivity.

2.5.4 1D and 2D Metal Oxides

Now that we have gathered an idea about the sensing mechanism involved in these one-dimensional nanostructures, let us look at some specific metal oxides like ZnO, SnO_2, TiO_2, In_2O_3, WO_3 and MoO_3 as case studies. The case studies presented below are very comprehensive and include all reports of 1D nano metal oxides that have been used for gas sensing applications

2.5.4.1 Zinc Oxide

Among the binary oxides that are to be discussed in this section, ZnO is probably the only metal oxide that possesses a rich family of one dimensional nanostructures — nanowires, nanobelts, nanorings, nanobows, nanocombs and a variety of other novel and unique architectures (Wang, 2004; Tian *et al.*, 2003). Zinc oxide belongs to the wurtzite family and is the most ionic among the oxides crystallizing in this structure (Heinrich, 1985). It has zinc and oxygen atoms in a tetrahedral coordination with the layers of anions and cations stacked alternatively one on top of the other. Thus anisotropic growth is favored in several directions, including $<2110>$, $<1010>$, $<0001>$. Thus in zinc oxide formation of such one or two-dimensional structures with the above planes as the exposure faces is energetically favored. It is a wide band gap (3.37eV) semiconductor with a high binding energy of ~ 60 meV. Due to its non-central symmetry, it exhibits piezoelectricity that makes it viable for using in actuators and transducers (Wang and Song, 2006). ZnO also exhibits near ultraviolet emission (Cross *et al.*, 2005) and transparent conductivity (Banerjee and Chattopadhyay, 2005). With this wide range of functional properties and the ability to be morphologically tailored to unique and novel shapes, ZnO is probably one of a kind.

The nanostructures will be classified based on the route of fabrication; for example ZnO nanostructures that have been grown from the vapor phase either by the vapor-liquid-solid (VLS) growth or vapor-solid growth (VS), those that have been synthesized using the chemical methods such as sol-gel, hydrothermal methods and so forth. The influence of oxide and metallic dopants on the gas sensing properties will also be touched upon. Wherever applicable, the sensing mechanism will be detailed.

a. Growth from the Vapor Phase   This route of fabrication is probably the most commonly employed for the fabrication of one dimensional nanostructures, as shown in Fig. 2.15. Usually growth from the vapor phase occurs via either vapor-solid or vapor-liquid solid route. It involves the thermal evaporation of a precursor material, either the metal oxide or the metal (in which case evaporation is followed by subsequent oxidation), in the presence of a carrier gas, with or without the influence of pressure. The carrier gas serves the purpose of transporting the species in the vapor phase to the substrate of choice and may or may not aid in oxidation. With control over the supersaturation factor high yields of 1-D nanostructures can be obtained.

Probably the first report on the gas sensing properties of ZnO nanowires was by Wan *et al.* (2004;a,b). In this study the ZnO nanowires were prepared by

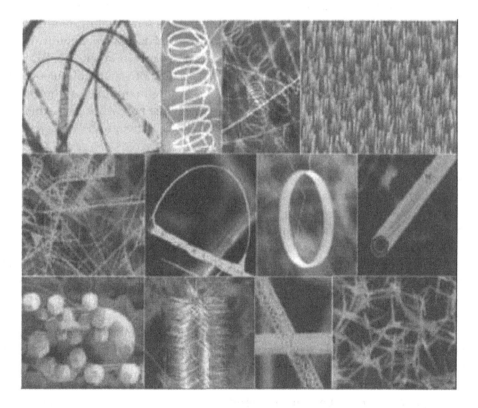

Figure 2.15.   ZnO nanostructures synthesized under controlled conditions by thermal evaporation of solid powders. [Reprinted from Materials Today, 6, Z. L. Wang, Nanostructures of Zinc Oxides, p.26, © (2004), with permission from Elsevier.] For color reference, turn to page 147.

evaporating Zn pellets at 900°C in an argon ambient, with a cross flow of oxygen. It was found that in this case controlling the oxygen concentration was the key to achieving massive quantities of ZnO. The nanowire sensors were evaluated for their sensitivity towards ethanol up to 1 ppm. The sensitivity at 1 ppm ethanol at an operating temperature of 300°C was 1.9 and increased with increase in ethanol concentration with a maximum sensitivity of about 47 at 200 ppm of ethanol. The mechanism of gas sensing was explained by means of the formation of oxygen traps (O_2^- ion) in ambient due to chemisorption and a resultant increase in surface resistance. On exposure to a reducing gas the electrons are returned to the surface thereby reducing the surface resistance. The high levels of sensitivity are attributed to the dimensions of the nanowire (~25 nm in diameter), since most of the atoms are now surface atoms. Till date this explanation is considered to be the closest approximation to what really happens on the surface of a nanowire and empirical studies are also being carried out simultaneously (Comini *et al.*, 2004).

Wan *et al.* (2004;c) in another set of experiments explain the formation and gas sensing characteristics of Cd doped ZnO to relative humidity. The nanowire network was synthesized by co-evaporating metallic Cd and Zn at 900°C in a tube furnace under 95%Ar and 5% O_2. The nanowires/belts had a narrow diameter distribution range and were about 2 μm in length. The sensor fabricated from the Cd-doped nanowires was tested for its sensitivity towards relative humidity (RH). The sensor demonstrated a three fold change in resistance at about 95% RH. The gas sensing mechanism is explained by the dissociation of water molecules at the nanowire tips and subsequent proton hopping at low humidities and just electrolytic conduction at high humidities. Interestingly the nanowires also exhibited a positive temperature coefficient of resistance (PTCR), though the exact mechanism for this unusual effect is not known.

Gao *et al.* (2005) have managed to produce ZnO/CdS core-shell like structures by initial thermal oxidation of ZnO/Zn particles at a temperature of 920°C that results in nanorods that are 20–40 nm in diameter and several microns in length, followed by ultrasonic coating of the nanorods with CdS particles. Ethanol sensing characteristics of the uncoated and coated ZnO nanorods were investigated and improvement in the sensing properties due to CdS doping has been reported.

ZnO nanobelt arrays have been fabricated by Wen *et al.* (2005) by controlled thermal oxidation of zinc foils. The observed nanobelts typically grow up to 15–20 μm in length and have widths in the range of 50–300 nm. They have been tested for their NH_3 sensing properties and good sensitivities have been reported.

Interconnected three dimensional networks of ZnO nanorods and nanowires were produced (Gao *et al.*, 2005) using a high temperature vapor-solid deposition process, by evaporating commercial ZnO powder at about 1400°C in the presence of argon as the carrier gas, under a constant pressure of about ~300 mbar.

A special mention about the multipods: Although the multipods are also usually obtained from vapor phase growth, their unique shape calls for a special mention under a separate heading. Multipods are characterized by a number of nanorods that are united at a common junction. The most common shape that has been reported so far is the tetrapod that has four nanorods emanating from a common junction (Gao and Wang, 2005; Wu *et al.*, 2005; Zhu *et al.*, 2005; Chu *et al.*, 2005). Inherent differences in the surface energies between the zinc and oxygen surfaces of the wurtzite structure are thought to drive the growth mechanism of these multipods.

b. Physical and Chemical Vapor Deposition Techniques Although physical and chemical vapor deposition techniques involve growth from the vapor phase, they are grouped under a different classification because they differ slightly from the usual evaporation methods. Chemical vapor deposition techniques can be used to synthesize large area arrays of nanowires by exercising a careful control over the supersaturation factor. Jeong *et al.* (2006) have fabricated a three dimensional hybrid nanostructure composed of a bottom ZnO layer from which an array of aligned ZnO nanowires are grown that has been used for oxygen sensing at room temperature. Liu et al (Liu *et al.*, 2005) used a combustion CVD process to synthesize single crystal flakes of wurtzite ZnO on a Si substrate. Ethanol sensing characteristics were investigated for a single crystal flake that revealed sensitivities of around 7.8 ($\Delta R/R_0$) for about 300 ppm ethanol, at an operating temperature of 400°C.

By utilizing magnetron sputtering and a series of controlled post-deposition annealing treatments Sreenivas *et al.* (2005) have manufactured nanowhisker networks of ZnO and also studied its alcohol sensing properties. Another approach of PVD that is widely used is site selective molecular beam epitaxy (MBE) that involves nucleating nanorods/nanorods arrays on a substrate coated with a catalyst such as Au or Ag (Kang *et al.*, 2005). Platinum and palladium are known catalysts for promoting dissociation of molecular H_2 to the more reactive atomic form. These nanorods arrays have been examined for their room temperature H_2 and ozone sensing properties.

Growth by Chemical Methods Wet chemical methods and hydrothermal methods (Xu *et al.*, 2006; Wang *et al.*, 2006) employ a chemical reaction followed by a low temperature heat treatment for completion of the reaction to produce the metal oxide. One primary characteristic that is common to all the chemical methods is that the aspect ratio of the nanostructures obtained is smaller than that compared to those that are grown by high temperature methods. This may due the fact that at such low temperatures, the energy available is not enough to grow them to lengths greater than a micron to a couple of microns. The usual dimensions that have been reported so far are around 50–60 nm in diameter and 1–2 microns in length on an average. The gas sensing properties have been investigated for primarily reducing species such as ethanol, LPG, HCHO and also for H_2S and very good sensitivities for low concentrations of these analytes have been demonstrated.

Recently flower like zinc oxide nanostructures that exhibited contact controlled resistance have been produced by a low temperature aqueous method (Feng *et al.*, 2005). The contact controlled resistance model is a crossover between the depletion layer model (generally used for describing the sensing mechanism in nanostructures) and the grain boundary model (used for polycrystalline thick films). In this model the resistance at the contacts of the nanorods and the shape of the potential barrier change with the change in the ambient environment.

Rout *et al.* (2006) have employed a room temperature solid state transformation mechanism to synthesize zinc oxide nanorods. Choice and proper usage of a surfactant such as triethanolamine is essential for the formation of nanorod bundles.

d. Miscellaneous Methods   This section groups those methods and those nanostructures that do not fit exactly any of the above descriptions. Electrophoretic deposition on anodic alumina membrane templates (with pore sizes of about 20 nm) has been used by Rout *et al.* (2006) to yield aligned nanowire bundles. Pulsed laser deposition (PLD) has been identified as another technique to produce aligned arrays of nanostructures. Sreenivas *et al.* (2005) have fabricated well-aligned columns of ZnO nanorods using PLD. Up to 12 μm long aligned nanowires with a preferred c-axis orientation have been synthesized by this method.

2.5.4.2   Tin Oxide Nanostructures and their Applicability in Sensing

Tin dioxide finds use in three major applications (a) as a transparent conductor — has high electrical conductivity and transparency in the visible range of the electromagnetic spectrum (b) as a catalyst in oxidation (c) as a solid state gas sensor. The latter two applications are controlled by the properties of the surface while transparency is more so a bulk property. SnO_2 has a band gap of 3.6 eV and is most commonly found in the rutile crystal structure although a high pressure orthorhombic variant is also known (Suito *et al.*, 1975). Next to zinc oxide, tin oxide is another metal oxide that has been widely synthesized in to nanowires, nanobelts, nanorods, nanoribbons, nanodiskettes and a range of other one dimensional nanostructures (Z.L. Wang's group, 2002–2003). Again as in the section on zinc oxide, attention will be only on those one-dimensional nanostructures that have been explored so far for gas sensing applications and the structures will be discussed based on the synthesis route. Figure 2.16 contains some electron microscopy images of the configurations of SnO_2 one dimensional nanostructures synthesized so far.

The most common method of producing one dimensional nanostructure of tin oxide also is growth from the vapor phase such as thermally evaporating a tin (Chen *et al.*, 2006; Ying *et al.* 2004; 2005) or a tin oxide precursor (Comini *et al.*, 2002; 2005) and subsequently condensing from the vapor phase. SnO_2 is known to have very good sensing properties towards reducing gases and in particular ethanol. So almost all these reports are on the ethanol sensing characteristics of ethanol except for a few where the other gases tested have been CO, NO_2 (Comini *et al.*, 2002; Ramgir *et al.*, 2005).

Dopants can also be incorporated into the nanostructures by co-evaporating or co-depositing the dopant sources during the synthesis process. CdS, Ru, Sb, CuO,

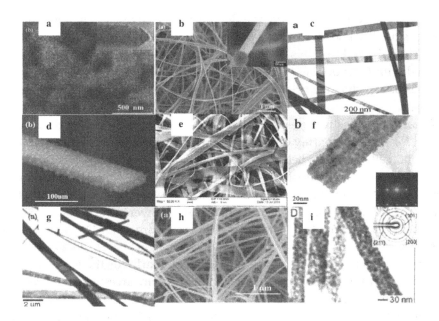

Figure 2.16. Some common morphologies of 1D SnO_2 nanostructures. [Figure 2.16(a) Reused with permission from Y. J. Chen, L. Nie, X. Y. Xue, Y. G. Wang and T. H. Wang, Applied Physics Letters, 88, 083105–1, (2006). © (2006), American Institute of Physics; (b). Reprinted from Nanotechnology, 15, Z. Ying, Q. Wan, Z. T. Song and S. L. Feng, SnO_2 nanowhiskers and their ethanol sensing properties, p. 1682, © 2004, with permission from Institute of Physics Publishing; (c). Reused with permission from E.Comini, G. Faglia, G. Sberveglieri, Z. Pan and Z. L. Wang, Applied Physics Letters, 81, 1869, (2002). © (2002), American Institute of Physics; (d). Reprinted from Materials Letters, 59, Z.Ying, Q. Wan, Z.T. Song and S.L. Feng, Controlled synthesis of branched SnO_2 whiskers, p. 1670, © (2005), with permission from Elsevier; (e). Reprinted from Sensors and Actuators B, 105, X. Kong and Y. Li, High sensitivity of CuO modified SnO_2 nanoribbons to H_2S at room temperature, p. 449, © (2005), with permission from Elsevier; (f) T. Gao and T. Wang, Sonochemical synthesis of SnO_2 nanobelt/CdS nanoparticles core-shell heterostuctures, Chemical Communications, 2004, p. 2558-Reproduced by permission of The Royal Society of Chemistry; (g). Reprinted from Sensors and Actuators B, 107, N.S. Ramgir, I.S. Mulla and K.P. Vijayamohan, A room temperature nitric oxide sensor actualized from Ru-doped SnO_2 nanowires, p. 708, © (2005), with permission from Elsevier; (h). Wan and T.H. Wang, Single-crystalline Sb-doped SnO_2 nanowires: synthesis and gas sensor application, Chemical Communications, 2005, p. 3841-Reproduced by permission of The Royal Society of Chemistry; (i). Reprinted with permission from Y. Wang, X. Jiang and Y. Xia, A solution-phase precursor route to polycrystalline SnO_2 nanowires that can be used for gas sensing under ambient conditions, Journal of the American Chemical Society 125, 2003, p. 16176. © (2003) American Chemical Society.]

Pd have been doped with SnO_2 nanobelts for improving the specificity towards a particular gas. CdS and Sb doped nanobelt heterostructures have been analyzed for their sensitivity towards ethanol and it was found that the doped sensors have a better sensitivity towards ethanol than the undoped ones. Ru has been envisaged as a dopant for a room temperature NO_2 and LPG nanobelts SnO_2 sensor. CuO on the other hand has been used as a dopant for selective room temperature H_2S sensing, since it is known to form a p-n junction with SnO_2, and the strong affinity of CuO to H_2S breaks up this p-n junction. Extremely high sensitivities have been achieved with CuO doping. The Pd decorated nanowires were seen to respond to pulses of H_2 while the pristine SnO_2 nanowires did not show any response to hydrogen.

Growth of nanostructures using templates has also been explored as a viable method for generating array of nanowires. For instance, porous aluminum oxide (PAO) and cellulose have been tried for templating SnO_2 nanowires (Kolmakov *et al.*, 2003). The method discussed in the latter involves growing metallic Sn nanowires in the pores of the alumina template, which is subsequently etched away. The Sn nanowires are then topotactically converted in to their oxide by controlled thermal oxidation. The nanowires grown are solid. On the other hand the nanostructures produced as discussed in (Hamaguchi *et al.*, 2006) are hollow and are the negative replicas of the template. Similarly for the template generated nanostructures the sensing properties are primarily investigated for reducing gases like ethanol, CO, H_2 and ethylene oxide.

Chemical synthetic methods such as hydrothermal route, aqueous synthesis have also been used for synthesis of these nanostructures (Wang *et al.*, 2003; Chen *et al.*, 2005). The chemical methods are usually carried out at low temperatures. These 1D structures have been investigated for their sensing properties towards ethanol, CO and H_2.

2.5.4.3 Indium Oxide

Indium trioxide is a transparent semiconductor that finds use in optoelectronic applications (Ishibashi *et al.*, 1990; Chopra *et al.*, 1983), and also in flat panel displays due to its high transparency and electrical conductivity (Jarzebski, 1982). Also it interestingly exhibits insulator-superconductor transition at low temperatures (Grantmakher *et al.*, 1998). Apart from this interesting array of properties In_2O_3 also is n-type semiconductor (owing to its oxygen vacancy doping (Bellingham *et al.*, 1991) that has been used as both nanostructured thin films and one-dimensional nanostructure for resistive gas sensing (Gurlo *et al.*, 1997;a;b)especially for oxidizing gases like NO_2, O_3. It also serves as a dopant in sensing films for improving the selectivity of sensors (Francioso *et al.*, 2006).

One dimensional In_2O_3 nanostructures, see Fig. 2.17, for gas sensing applications have been primarily produced by laser ablation method using Au as catalyst on a Si/SiO_2 substrate. The In_2O_3 nanostructures so produced were found to have a cubic crystal structure with [1 1 0] as their primary growth direction (Li *et al.*, 2003). They have been primarily examined for their gas sensing properties towards NO_2, NH_3. Other VOCs tested include butylamine and ethanol for sensing using In_2O_3.

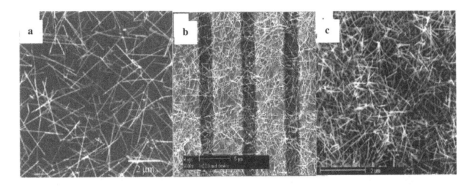

Figure 2.17. One-dimensional In$_2$O$_3$ synthesized by the laser ablation method. [Fig. 2.17(a). Reprinted with permission from C. Li, D. Zhang, B. Lei, S. Han, X. Liu and C. Zhou, Surface treatment and doping dependence of In$_2$O$_3$ nanowires as ammonia sensors, *Journal of Physical Chemistry B*. 107, 2003, p. 12451. © (2003) American Chemical Society; (b). Reprinted with permission from D. Zhang, Z. Liu, C. Li, T. Tang, X. Liu, S. Han, B. Lei and C. Zhou, Detection of NO$_2$ down to ppb levels using individual and multiple In$_2$O$_3$ nanowire devices, *Nano Letters* 4, 2004, p. 1919–1924. © (2004) American Chemical Society. (c). Reprinted from the Annals of the New York Academy of Sciences, 1006, C. Li, D. Zhang, S. Han, X. Liu, T. Tang, B. Lei, Z. Liu and C. Zhou, Synthesis, applications and electronic properties of indium oxide nanowires, p.104, with permission from Blackwell Publishing.]

2.5.4.4 Tungsten Oxide

WO$_3$ is a well-known n-type semiconductor with a band gap of 2.6 eV that has been used not only in solid-state gas sensors but also in catalytic/photocatalytic (Lietti *et al.*, 1996) and electrochromic applications (Livage and Guzman, 1996). WO$_3$ exhibits a rich variety of polymorphic transformations and exists in triclinic, monoclinic, orthorhombic, tetragonal and in some cases the metastable hexagonal form. The crystal structure is a variation of the cubic perovskite structure, with octahedrally coordinated metal ions. Electrical properties are governed by oxygen vacancy mediated processes wherein oxygen vacancies in the parent lattice act as donor states and contribute to electrical conductivity.

Although abundant research has been done on gas sensing properties of nanostructured WO$_3$ thin films synthesized by various routes, there are not many reports on one-dimensional nanostructures of WO$_3$ used for gas sensing. Fig. 2.18 shows scanning electron micrographs of one-dimensional structures of WO$_3$ synthesized by different routes. In the approach described by Parthangal *et al.* (2005) a previously deposited tungsten film is restructured in to nanowires by heating the substrate in an inert gas CVD chamber. Although the WO$_3$ nanowires are non-stoichiometric good sensitivities to NO$_2$ are obtained. Wet chemistry method employed by Polleux *et al.* (2006) has yielded porous nanobundles of WO$_3$ nanofibers that exhibited excellent sensitivities down to ppb levels of NO$_2$. Similar porous WO$_3$ nanowire mats have also been produced by a colloidal approach. Very good sensing responses towards NO$_2$ and NH$_3$ have been achieved using this method (Kim *et al.*, 2005).

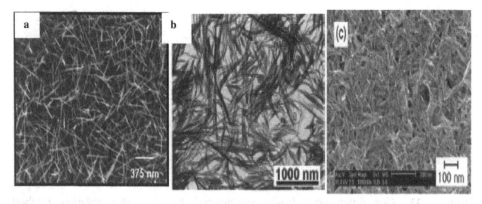

Figure 2.18. One dimensional structures of WO$_3$ synthesized by different routes. [Fig. 2.18(a). Reprinted from Journal of Materials Research, P.M. Parthangal, R.E. Cavicchi, C.B. Montgomery, S. Turner and M.R. Zachariah, Restructuring tungsten thin films in to nanowires and hollow square cross-section microducts, 20, p. 2889, © (2005), with permission from Materials Research Society; (b) Reprinted from Angewandte Chemie, 45, J. Polleux, A. Gurlo, N. Barsan, U. Weimar, M. Antonietti and M. Niederberger, Template-free synthesis and assembly of single-crystalline tungsten oxide nanowires and their gas sensing properties, © 2006, with permission from Wiley-VCH Verlag GmbH & Co KG; (c). Reused with permission from Y.S. Kim, S.C. Ha, K. Kim, H. Yang, S.Y. Choi, Y.T. Kim, Applied Physics Letters, 86, 213105–1, 2005. © (2005), American Institute of Physics.]

2.5.4.5 Molybdenum Oxide

Molybdenum oxide contains a family of oxides that form in various crystal structures such as orthorhombic, monoclinic, hexagonal, tetragonal etc. and also incorporate the well-known 'Magneli' (MoO_{3-x}) phases. Among these, molybdenum trioxide is an n-type semiconductor wherein the conductivity is a function of lattice oxygen deficiencies and has a band gap of 3.2eV. Molybdenum trioxide is an interesting oxide in that it has a modified cubic ReO$_3$ structure and the orthorhombic form of MoO$_3$ has a unique layered morphology with distorted edge-sharing octahedra. Octahedral layers share edges along the a-axis [1 0 0] and corners along the c-axis [0 0 1]. To the best of our knowledge, this is the first report of single crystalline one-dimensional MoO$_3$ nanostructures for use in gas sensing (Gouma *et al.*, 2006), although there have been reports of MoO$_3$ nanorods (Comini *et al.*, 2005; Taurino *et al.*, 2006) that have been used for sensing NO$_2$, NH$_3$, CO, CH$_3$OH. The MoO$_3$nanorods as reported in (Comini *et al.*, 2005) were manufactured by sublimation and condensation of a Mo foil on an alumina substrate while the ones in (Taurino *et al.*, 2006) have been synthesized by template directed hydrothermal synthesis.

The author's group has used a novel approach (Gouma *et al.*, 2005; Sawicka *et al.*, 2005) to produce *single crystalline*, one-dimensional nanowires. It is a single step process that makes use of electrospinning to produce a polymer/metal oxide nanocomposite fibrous mat. Figure 2.19 shows a typical HRTEM image of a MoO$_3$ nanowire (inset shows a magnified image of the nanowire) and the sensing response of the nanowire to increasing concentrations of NH_3. The nanocomposite is then heat treated at an elevated temperature to attain crystallinity and also to burn of the polymer. The metal oxide nanowires obtained are single crystalline and have a uniform oxide polymorph distribution. The exact mechanism of formation of these nanowires is not known, but the polymer nanofibers somehow act as a template for the growth of single crystal nanowires. The author's research has also produced WO_3 and MoO_3 nanowires and used them for sensing NO_2 in the case of WO_3 and NH_3 in the case of MoO_3. The nanowire sensors showed excellent sensitivity (measured as R_g/R_{air}), compared to the sol-gel nanostructured thin films.

2.5.4.6 Vanadium pentoxide

Vanadium pentoxide is the common phase form that is used in gas sensing among the different phases of vanadium oxide. It possesses a layered orthorhombic crystal structure and has an optical band gap of 2eV (Surnev *et al.*, 2003). The layered structure enables the material to retain a wide variety of other molecules or cations to be embedded in between these layers. When aqueous routes are used to synthesize these nanostructures (which are the most common methods used) they may retain water ($V_2O_5.xH_2O$) between the layers of their crystal structure.

V_2O_5 nanobelts, as shown in Fig. 2.20, have been synthesized using a mild hydrothermal treatment with high yield (Liu *et al.*, 2005). These nanobelts were found to be highly selective towards ethanol up to concentrations of about 10ppm. Coating these nanowires with different metal oxides such as Fe_2O_3, TiO_2, SnO_2 the sensitivity towards alcohol was found to increase than in the uncoated nanobelts, and the increase in sensitivity is maximum for those sensors coated with SnO_2. Helium has also been detected using V_2O_5 nanowires synthesized using the polycondensation of vanadic acid (Yu *et al.*, 2005). Polycondensation of metal vanadate ions also is known to yield nanofibers several microns long (Raibler *et al.*, 2005) although this is a very time-consuming procedure. Such nanofibers have been reported to possess extremely high sensitivities towards very low concentrations of amines ($\sim$30ppb limit of detection).

2.6 FUTURE CHALLENGES

Preliminary experiments with one-dimensional nanostructures done so far have shown that they possess enhanced properties for the desired applications. But the ultimate challenge will be to fabricate single nanowire devices, in order to realize their potential to the fullest extent. Isolation of a single nanowire from a mat of nanowires or separation of a single nanorod from an array of nanorods requires the use of nanomanipulation techniques. For laboratory scale experiments, usage of these techniques is justified. But on the industrial scale not only are they complicated, they are also difficult to be made viable for mass production of single nanowire devices.

The next level of difficulty lies in making the transition from actual 'nano' structures to 'micro' scaled devices and electrical connections for interfacing with

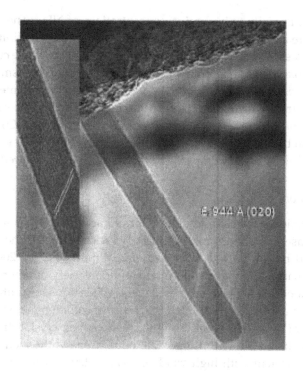

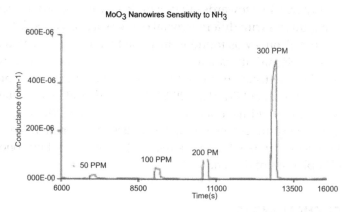

Figure 2.19. (a) HRTEM image of a MoO_3 nanowire; inset shows a magnified image of the nanowire. (b) Sensing response of the nanowire to increasing concentrations of NH_3.

real world systems. Manufacturing substrates with the appropriate electrodes (in the case of sensing) for both sensing and heating these devices still continues to be a challenge. Although direct deposition of heater and sensing electrodes using vapor phase methods has been tried, possible contamination to the actual sensing matrix is a concern. Once these challenges are overcome, one-dimensional nanostructures can tremendously improve the sensor sensitivity, selectivity for detection of single molecules and pico grams of analytes.

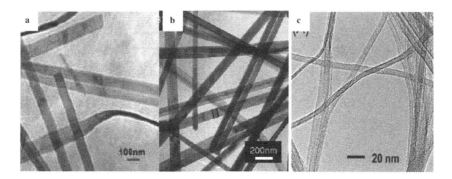

Figure 2.20. TEM images of V$_2$O$_5$ nanobelts. [Fig. 2.20(a). Reprinted from Advanced materials, 17, J. Liu, X. Wang, Q. Peng and Y. Li, Vanadium pentoxide nanobelts: highly selective and stable ethanol sensor materials, p.764, © 2005, with permission from Wiley-VCH Verlag GmbH & Co KG;(b). Reprinted from Sensors and Actuators B, 115, J. Liu, X. Wang, Q. Peng and Y. Li, Preparation and gas sensing properties of vanadium oxide nanobelts coated with semiconductor oxides, p. 481, © 2006, with permission from Elsevier. (c). Reprinted from Sensors and Actuators B, 106, I. Raible, M. Burghard, U. Schlecht, A. Yasuda, T. Vossmeyer, V2O5 nanofibers: novel gas sensors with extremely high sensitivity and sensitivity to amines, p. 730, © 2005, with permission from Elsevier.]

SUMMARY

Nanomaterials have the potential for revolutionizing the sensing field, with the development of one-dimensional nanostructures that provide the maximum surface area to volume ratio. Also the unconventional nanoassemblies will result in the development of miniaturized sensor arrays to be used in electronic noses aiding in the simultaneous detection of multiple analytes. The future lies in the successful integration of these nanoassemblies and arrays with electronics for applications ranging from molecular recognition to bio sensing.

Chapter Three

Hybrid Nanomaterials

3.1 BIO-DOPED OXIDE NANOSENSORS

The term bio-doped ceramic encompasses the entrapment of any biological entity like biomolecules, cell, bacteria, and viruses, within a given ceramic matrix resulting in added functionality. Biomolecules are naturally occurring molecules in living organisms, e.g. amino acids and nucleotides, consisting primarily of carbon, hydrogen, oxygen, nitrogen, phosphorous and sulfur. Hierarchical biomolecule assembly is the basis of life on earth. Ceramic materials and metal oxides in particular, are usually non-toxic, inert, and chemically and thermally stable and so they may be used in applications where biocompatibility and/or thermal stability requirements are essential. The synergistic combination of biological entities and ceramics has lead to an exciting new class of materials that may serve as the core components of the next generation biosensors, enzyme reactors, or controlled release agents. The performance, functionality and application of such bio-doped ceramics will depend on the activity, functionality and interaction of entrapped biomolecules with the surrounding ceramic matrix. It is therefore of prime importance to understand how biological entities, such as proteins, interact with a porous ceramic matrix such as silica sol-gels. Bioencapsulation is a very active field of research with numerous publications to date and various reviews. Recently, biodoped ceramics have been reviewed with a focus on their applications (Avnir *et al.*, 2006; Sanchez *et al.*, 2005).

Focusing on hybrids in which biologicals are encapsulated within a metal oxide matrix this review attempts to address the theory of encapsulation process and the studied interactions between the organic component ("dopant") and the inorganic host. Biologicals are readily dispersed in solutions and are known to have complete freedom of movement. Encapsulation or entrapment is a technique that restricts such freedom. It restricts their movement in space either completely or confines it to a small limited region. It may increase the stability of enzymes and also improve their reaction kinetics. The requirements for effective and useful bio-encapsulation include a high density of entrapped biological, high bio-activity, long-term stability under potentially adverse reaction conditions, good accessibility to analytes, rapid response times to any analytes of interest and resistance

Nanomaterials for Chemical Sensors and Biotechnology by Pelagia-Irene (Perena) Gouma
Copyright © 2010 by Pan Stanford Publishing Pte Ltd
www.panstanford.com
978-981-4267-11-3

to leaching and/or desorption. In this treatment of bio-encapsulation within a ceramic matrix, the effect of entrapment on various biological entities (enzymes, bacteria, and cells), the interactions between host matrix and encapsulated entity, as well as some key applications of biodoped ceramics will be discussed.

3.2 BIO-DOPED CERAMIC SYNTHESIS

Nature has taught us of numerous biospecies-ceramics interactions that have been reproduced in the lab. Silica is produced biologically by horsetail, rice husk, diatoms, or marine sponges. Silicatein, a protein, has been recognized to direct the assembly of biologically synthesized silica (Cha *et al.*, 1999; Adamson *et al.*, 2004; Belcher *et al.*, 1998). Silica and titania particles have been synthesized using a fungus *Fusarium oxysporum* (Bansal *et al.*, 2005). Zirconia particles have also been synthesized by fungus extra cellular proteins (Bansal *et al.*, 2004). Virus particles have been used as templates for the synthesis of inorganic materials such as tungsten trioxide (WO_3) and molybdenum trioxide (MoO_3)(Douglas and Young, 1999). Recently, advances in soft chemistry routes, and sol-gel technology in particular, have allowed encapsulation of various enzymes, proteins, cells, and bacteria within the ceramic matrix (Wei *et al.*, 2002; Kadkinova and Kostic, 2001). This section describes common synthesis routes for bio-doped ceramic synthesis.

3.2.1 Sol-Gel Technique for Ceramic Processing

The sol-gel technique is used to synthesize ceramics at room temperature and pressure conditions. It typically involves the formation of 'sol' using a suitable precursor and solvent, and its subsequent gelation. 'Sol' is a colloidal suspension of metalorganic precursors. Colloidal particles can be defined as particles having at least one dimension in the range $20\,\text{Å} - 0.2\,\mu\text{m}$ and containing $10^3 - 10^9$ atoms. A 'Sol' is an intermediate between a true molecular solution and coarse particle dispersion (Livage and Lemerle, 1982). The mean particle size in colloidal suspension is extremely small so any gravitational forces acting on these particles are considered to be negligible. Thus, short range forces like Van der Waal's forces control the interaction between the particles. Metal alkoxide precursors are metalorganic and they have an organic ligand attached to the metal or metalloid atom, e.g. Tetraethyl Orthosilicate (TEOS) $Si(OC_2H_5)_4$, Tetramethyl Orthosilicate (TMOS), and $Si(OCH_3)_4$. Metal alkoxides undergo hydrolysis upon contact with water. During hydrolysis a hydroxyl ion attaches to the metal atom and alcohol is the byproduct of the reaction (equation 3.1).

$$Si(OR)_4 + H_2O \longrightarrow HO - Si(OR)_3 + R - OH \qquad (3.1)$$

Hydrolysis is typically carried out in the presence of a catalyst. Alcohol is added as co-solvent as it is a common solvent to both the metal alkoxide and the water. In the presence of sufficient water and catalyst, hydrolysis can go to completion as follows (equation 3.2),

$$Si(OR)_4 + 4H_2O \longrightarrow Si(OH)_4 + 4R-OH \tag{3.2}$$

The condensation reaction follows hydrolysis, where two partially hydrolyzed silanol groups condense to form a siloxane group. Either water or alcohol is released as a by product of condensation reaction (equations 3.3 and 3.4 respectively)

$$(OR)_3Si-OH + HO-Si(OR)_3 \longrightarrow (OR)_3Si-O-Si(OR)_3 + H_2O \tag{3.3}$$

or

$$(OR)_3Si-OR + HO-Si(OR)3 \longrightarrow (OR)_3Si-O-Si(OR)_3 + ROH \tag{3.4}$$

As the condensation reaction continues, larger aggregates are formed and the viscosity of the sol continues to increase and subsequently a gel is formed. The gelation point is described by the percolation theory as the condition at which the first single cluster of molecules that spans the entire sol appears (Brinker and Scherer, 1990). A gel consists of interpenetrating continuous solid and liquid phases. The collapse of the solid phase is prevented by the presence of the liquid phase. The solid phase on the other hand confers structural stability to the liquid phase and prevents it from flowing out, thus a self-standing structure is formed. The continuous solid network confers elastic properties to a gel. The condensation process does not stop after the gelation point is reached. It continues beyond it, attaching smaller aggregates to the sample-spanning network. This continued condensation is called aging of gels. These 'wet' gels can be dried to remove solvent from the pores of gel. A gel dried under ambient conditions is termed 'xerogel'. Drying has a profound effect on the microstructure of the xerogel. Capillary forces arising due to solvent removal from the pores lead to shrinkage of the gel (Smith *et al.*, 1995). Gels can also be dried supercritically, in which case shrinkage can be avoided. Supercritically dried gels are called aerogels.

As sol-gel is a solution based process, it is adaptable to produce materials in various forms such as thin films, fibers, disks, and other bulk shapes. It has the unique advantage of ambient temperature processing; moreover the molecular level mixing of constituents allowed during sol-gel synthesis offers great control over the composition and properties for bio-doped ceramics. Silica sol-gel, in particular, exhibits optical transparency and bio compatibility making it a very attractive encapsulation matrix for biological such as proteins, cells etc. Various processing parameters such as the choice of precursor material, catalyst type (acid/base), and the extent of hydrolysis control the final microstructure of the gel. This means that tailored silica gel properties can be obtained by choosing the appropriate processing parameters, as discussed below. Although almost all metal oxides can be synthesized by means of sol-gel processing, their synthesis conditions are not always bio-compatible. In the case of transition metal alkoxides, the hydrolysis and condensation reactions are very rapid. Thus, the processing window for these metal oxides is very narrow. Their chemistry is not as flexible as silica due to their intrinsic properties (Pierre, 2004).

3.2.1.1 Modified Sol-Gel for Bioencapsulation

Traditional silica precursors such as TEOS and TMOS can also be modified by incorporation of an organic functional group. The organic group is attached to an inorganic siloxane precursor through a Si-C bond. Such organically modified silica glasses are called 'ormosils'. By choosing an organic group having the properties of ormosils such as wettability, both ionic or hydrogen bond formation capacities and porosity can be tuned. Reactive functional groups of such modified precursor can covalently bond with either the metal oxide or the biological entity. However ormosils are generally less robust than inorganic glasses (Böttcher *et al.*, 2004; Gill, 2001; Tripathy, 2001). The better biocompatibility of ormosils is advantageous over traditional silica sol-gel. For example, the encapsulation of lipase within an ormosil resulted in an increased stability and a high relative enzyme activity (Reetz *et al.*, 1996). Interactions between hydrophobic regions of organically modified silica and lipase may have contributed to this increased stability of lipase. Commonly used ormosil monomers are shown in Fig. 3.1.

One of the important requirements for successful bioencapsulation is that the biological entity retains its functionality following encapsulation. Biological entities are very sensitive to their surrounding environment and their functionality depends on various factors such as their shape, surface charges, the presence

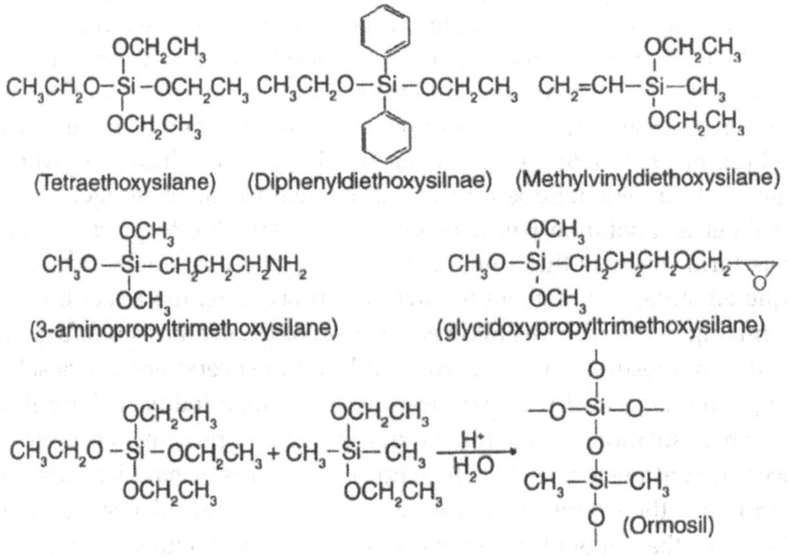

Figure 3.1. Tetraethyl orthosilicate and commonly used ormosil monomers and reaction scheme for ormosil formation.[Reprinted from Tripathi *et al.*, Copyright (2005), with permission from Elsevier.]

of inhibitor molecules, temperature and pH of their surrounding media among others. Though the sol-gel route offers ambient temperature conditions for the encapsulation of biological entities, alcohol produced during the hydrolysis and condensation steps and extreme pH conditions can have a detrimental effect on the functionality of biological entities. Alcohol or extreme pH values may lead to denaturation and/or aggregation of proteins; alcohol is also toxic to the living cells (i.e. cytotoxic). Alcohol produced during condensation of TMOS led to denaturation of atrazine-degrading enzymes (Kauffmann and Mandelbaum, 1996).

Ellerby *et al.*, 1998, modified the sol-gel procedure to avoid the adverse effects of pH and alcohol on enzyme stability. Their strategy was two fold: first, they avoided the addition of excess alcohol to the sol; they ensured better distribution of precursor within the solvent by sonicating the sol before the addition of the protein solution, and second: to avoid the adverse effect of extreme pH due to the catalyst used during the hydrolysis of the sol, they added buffer to the sol just before adding the protein solution. Addition of the buffer pulled the pH of sol to 5. They successfully encapsulated bovine copper-zinc super oxide dismutase (CuZnSOD), horse heart cytochrome C, and horse heart myoglobin (Mb) in sol-gel silica matrices. Another simple method to get rid of alcohol during sol-gel encapsulation is to evaporate alcohol from the sol before the addition of biological entity. Ferrer *et al.*, 2002, eliminated the alcohol generated in hydrolysis of TEOS by gentle vaccum rotavaporization before adding the protein solution in buffer to the sol. Horseradish peroxide (HRP) was encapsulated in silica gel using this alcohol-free route. Its activity was completely preserved upon encapsulation.

Ferrer *et al.*, 2003, compared the effects of alcohol evaporation route on the viability of *Escherichia coli (E. coli)* in pure and in hybrid silica matrix, incorporating poly (ethylene) glycol (PEG) and glycidoxypropyltrimethoxysilane (GPTMS). It was shown that the alcohol level within the silica gel affects the viability of bacteria most significantly. The removal of alcohol by evaporation leads to better viability of bacteria after encapsulation. Furthermore, the hybrid silica matrix preserved the viability of the bacteria for a longer time than what the pure silica matrix did. The use of sodium silicate as a precursor for silica gel also prevents the addition /formation of alcohol during the sol-gel encapsulation of biological entities. In conventional sol-gel synthesis of silica using sodium silicate as a precursor, acid is added to sodium silicate solution in water. This produces silisic acid, further polymerization of silicic acid leads to formation of silica gel. Bhatia and Brinker, 2000, modified conventional sodium silicate processing to accommodate biological entities within the sol-gel. A pH-adjusting buffer was added to sodium silicate sol before the addition of protein solution. Using this method, they were able to retain 73% of the activity of horseradish peroxidase (HRP) against that of free HRP. Na^+ ions produced during hydrolysis of sodium silicate need to be removed using acid action exchange resin as they are likely to have adverse effects on the biological entities beyond a critical concentration. Coeffier *et al.*, 2001, compared the effects of sol-gel encapsulation using TEOS as silica precursor and using sodium silicate as silica gel precursor on the viability of encapsulated bacteria *Escherichia*

Coli. It was found that aqueous sol-gel encapsulation using sodium silicate better preserves the viability of encapsulated bacteria than alkoxide based sol-gel encapsulation using TEOS as a precursor. In these studies the absence of alcohol during sodium silicate sol-gel encapsulation helped bacteria preserve their viability.

Apart from removing toxic alcohol from sol-gel matrix, the stability of the encapsulated biological entity can be improved by the addition of amino acids, sugars, or cyto-protecting agents such as glycerol to the sol-gel. The thermal stability and activity of silica-sarcosine-encapsulated α-chymotropsin was better than only TEOS-encapsulated α-chymotropsin (Brennan *et al.*, 2003). Pin-printed protein micro arrays were successfully fabricated using a modified sol-gel precursor such as monosorbitol silane (MSS) and diglyceryl silane (DGS). DGS and MSS release glycerol and sorbitol, respectively as byproducts of the initial hydrolysis reaction during their sol-gel processing. Glycerol or sorbito are thus introduced into the sol-gel matrix upon polycondensation. These protein stabilizers help encapsulated protein in retaining its activity as opposed to TEOS sol-gel process that releases harmful ethanol during hydrolysis (Rupcich *et al.*, 2003). It was observed that encapsulation of enzymes Factor Xa, dihydrofolate reductase, cyclooxygenase-2, and ς-glutamyl transpeptidase within sol-gel using DGS as precursor, resulted in increased stability and activity than within the sol gel using TEOS as a precursor. This increased stability and activity was attributed to the release of gycerol from DGS (Besanger *et al.*, 2003).

Extensive shrinkage observed in silica gels prepared from TEOS upon drying can lead to matrix cracking and subsequent denaturation of encapsulated protein. TEOS monoliths showed 85% shrinkage after air for 60 days, whereas DSG- and MSS- derived gels showed 65% and 50% shrinkage under the same conditions. TEOS also showed higher rate of shrinkage (Brook *et al.*, 2004). The lower shrinkage of DSG- or MSS-derived gels can be due to slow evaporation of glycerol or sorbitol during drying than ethanol. Slow evaporation liquid phase prevents significant pore collapse and thus less shrinkage (Chen *et al.*, 2005). Further significant reduction in the shrinkage of DSG-derived gels was observed within gels derived from the co-hydrolysis of DGS with modified trifunctional silanes gluconamidylsilane (GLS) and the analogous maltonamidylsilane (MLS). DSG-GLS-derived glasses showed 10% shrinkage, significantly lower than either TEOS of DGS alone. Sugars associated with GLS or MLS are covalently linked into the matrix. Water layer hydrating these sugars confer plastic properties to the matrix. This hydrated layer of sugars resists the capillary forces that accompany drying. Decreased surface area upon addition of GLS or MLS to DGS results in loss of micropores that can also lead to a significant decrease in capillary forces and subsequent shrinkage. Addition of sugar GLS to a DGS-based encapsulation matrix led to increased stability of HRP as compared with an unmodified DGS-based encapsulation matrix. The timing of addition of GLS also affected stability of encapsulated HRP. When GLS was added to preformed DGS monolith it increased 3-fold and when GLS was added to DGS sol it increased 4-fold as compared with an unmodified DGS (Sui *et al.*, 2006).

Polymer additives can also be used for better stability of protein upon encapsulation. The activity of Lipase in TEOS-derived silica increased with addition of increasing concentrations of PEG. PEG also was found to confer crack-resistant properties to the silica matrix. PEG destabilizes the unfolded configuration of enzyme by reducing its solubility in water and hence the protein tends to remain folded in the presence of PEG. However Polyvinyl acetate (PVA) added to TEOS sol-gel reduced the activity of lipase encapsulated in silica gel (Keeling-Tucker *et al.*, 2000). Antibiotic solutions have also been added to the sol-gel matrix while encapsulating proteins. This practice reduces the risk of degradation of encapsulated proteins due to microbial attack (Luckarift *et al.*, 2004).

Goring and Brennan, 2002, prepared sol-gel-derived nanocomposites films suitable for protein encapsulation. These thin films were prepared using tetraethyl orthosilicate, methyltriethoxysilane and/or dimethyldimethoxysilane and PEG. Human serum albumin (HSA) was encapsulated within these sol-gel films. It was found that it retained at least partial activity. The improved crack resistance of the TEOS-derived films was attributed to greater hydration of the films during drying, which resulted in relatively less hydration stress during rehydration.

The sol-gel process has been modified for biological entities to be encapsulated within the sol-gel layer by Chemical Vapor Deposition (CVD). This process is also known as 'biosil' method, and uses a gaseous flux of alkoxysilane. A condensation reaction occurs on the surface of the moist biological entity, depositing a layer of silica gel on its surface. The advantages of this procedure include the immediate removal of the alcohol produced during the condensation and control over the thickness of the deposited silica layer. This method is very flexible and can be extended to various types of precursor solutions such as other metal oxides precursor or ormosils (Böttcher *et al.*, 2004; Caturan *et al.*, 20040. Horseradish peroxidase (HRP) was encapsulated in titanium sol-gel using this method. The HRP electrode was prepared by dropping protein solution onto the surface of a pretreated glassy carbon electrode. Titanium isopropoxide was absorbed on to the electrode by suspending it over titanium isopropoxide solution at 25°C for six hours. Hydrolysis of adsorbed titanium isopropoxide over HRP resulted in a titanium sol-gel coating over HRP. This film of titanium oxide prevented leaching of enzyme and retained its activity (Yu and Yu, 2002). Various cells, bacteria and enzymes have been successfully encapsulated by using the biosil method (Pressi *et al.*, 2003; Boninsegna *et al.*, 2003).

The freeze-gelation technique offers low cost processing of crack free and zero shrinkage ceramics (Statham *et al.*, 1998) so bio-doped ceramics may be processed by this method. The process involves mixing of SiO_2, ceramic powder and the biological component in an aqueous state. This slurry is solidified by a freezing process (freezing temperature $\sim -40°C$). This process of freezing causes irreversible gelation of the ceramic slurry. The shape attained due to freezing is retained even after the increase in temperature of the structure. The porosity induced within the freeze-dried gel depends on the ice-formation due to the freeze process. The freeze-gelation technique has been used for encapsulation of cells and

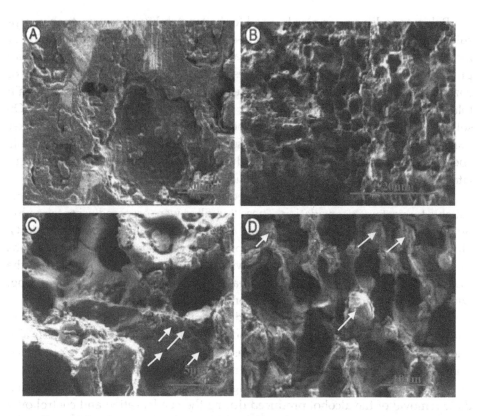

Figure 3.2. Scanning electron micrographs of *Bacillus Sphaericus* or *Saccharomyces Cerevisiae* (yeast) encapsulated within a ceramic by the freeze gelation method. A slurry of ceramic powders and biological component is solidified at freezing at very low temperatures. Thus-formed biodoped ceramic retains its shape even when the temperature becomes normal. (A) Surface of a *B. Sphaericus*-encapsulating ceramic. (B and C) Fracture surfaces of *B. Sphaericus* encapsulating ceramic. (D) Fracture surface of a *S. Cerevisiae*-encapsulating ceramic. Embedded cells are marked with arrows (Soltmann *et al.*, 2003). For color reference, turn to page 148.

spores of *Bacillus Sphaericus* and *Saccharomyces Cerevisiae* Fig. 3.2. The number of living cells decreased after freeze gelation and drying. However surviving cells showed good physiological activity and good storage life (Soltmann *et al.*, 2003).

3.2.2 Other Techniques for the Synthesis of Bio-Doped Ceramics

The broad size distribution of pores in conventional sol-gel material may be undesirable for applications such as biosensors. Ordered mesoporous silica structures offer a narrow pore size distribution. Yang *et al.*, 2006, encapsulated various enzymes in ordered mesostructured silica by the fish-in-net method as shown in Fig. 3.3. Here interactions between enzyme and preformed precursor lead to tailored cages within which enzymes were encapsulated. The loading of enzyme within the sol-gel materials plays an important role in determining the function

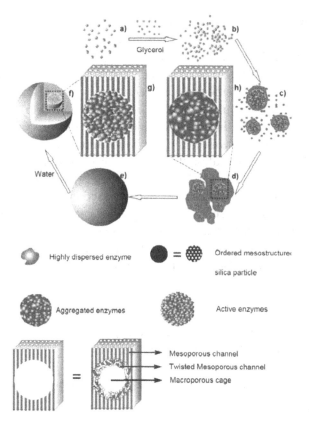

Figure 3.3. "Fish-in-Net" encapsulation of enzymes in macroporous cages as stable, reusable, and active heterogeneous biocatalysts. (A) Buffer solution containing highly dispersed enzymes. (B) Preformed precursor with ordered mesostructured silica particles mixed with enzyme. Tetraethyl orthosilicate was assembled from a triblock ethylene oxide(EO)/propylene oxide (PO) co-polymer surfactant ($EO_{20}PO_{70}EO_{20}$,P123) in ethanol. Evaporation of ethanol and addition of glycerol yielded these preformed precursors. (C) Assembly of enzymes with ordered mesostructured silica particles. (D) Interactions between enzyme and preformed precursor led to tailoring of pores. Enzymes were trapped in the "net" formed by polymerization and condensation of ordered mesostructured silica particles. (E) Formation of silica spheres encapsulating enzymes in microporous cages. (F) Silica spheres with encapsulated enzyme under biocatalytic conditions. When water is introduced into these cages, and the chemical environment is similar to that of a native enzyme in solution, it is possible that enzymes are dispersed. (G) Magnification of the enzyme-filled macroporous cage from (F). (H) Magnification of the enzyme-filled macroporous cage from (D) where enzymes are aggregated in macroporous cages. [Reprinted from Yang *et al.*, 2006, with permission from WILEY-VCH Verlag GmbH & Co.] For color reference, turn to page 149.

of devices such as biosensors. Narang *et al.*, 1994, reported sandwich immobilization of urease within two layers of sol-gel films. An appropriate TEOS sol was spin coated onto glass slides. These glass slides were then cured at 200°C for 10 minutes. A layer of urease solution was then deposited onto this sol-gel film. After storage at 4°C for 24 hours, another layer of TEOS sol-gel was spun on top of it. This method enabled high loading of enzyme within the ceramic matrix. Urease entrapped in this sandwich sol-gel film retained more than 95% activity for 6 weeks when it was stored at 4°C (Goring and Brennan, 2002). Ma et al., 2004, used a microemulsion technique to synthesize thin silica shells encapsulating single molecules of protein met-myoglobin (Mb). A solution of an anionic surfactant, sodium bis(2-ethylhexyl) sulfosuccinate (NaAOT), and isooctane was prepared. An oil/water emulsion was prepared by addition of Mb solution to NaAOT/isooctane solution (0.1 M). This protein containing microemulsion was then maintained at 304K. TEOS solution (0.01 M) was then added to it with shaking. The nano aggregates formed by the reaction of TEOS at the oil/water interface were 20–30 nm in size. These aggregates consisted of individual silica particles encapsulating Mb. The proteins wrapped in a thin silica shell preserved their structure and activity. In fact their stability was increased upon encapsulation. This encapsulation also protected the protein from leaching out (Ma et al., 2004).

3.3 ENTRAPMENT OF BIOLOGICALS WITHIN A CERAMIC MATRIX — THE EFFECTS ON PROPERTIES

3.3.1 Theory of Encapsulation of Biological Within Porous Matrices

Proteins are essential elements for the function of all living cells and viruses. They are polymers having a defined sequence of α-amino acids. There are twenty amino acids commonly found in proteins. All twenty amino acids, except proline, have a negatively charged carboxyl group and a positively charged amino group. The protein structure is a hierarchical structure consisting of four levels. The primary structure is a defined amino acid sequence formed by a covalent interaction between participating amino acids with the exception of disulfide bonds. Ordering of a primary structure into a helix or β-sheets via hydrogen bonding forms the secondary structure of the protein. The tertiary structure is a three-dimensional (3D) configuration arising due to the overall folding of the secondary structure. The latter folds in such a way that side chains of non-polar amino acids (hydrophobic) remain inside the 3D structure, whereas side chains of polar (hydrophilic) amino acids are exposed on the outer surface. The quarternary structure is the association of two or more tertiary structures. These structures are stabilized mainly by *non-covalent interactions* like *hydrogen bonding, Van der Waal's* interactions and *ionic bonding*. Proteins can have different functions like structural, mechanical, signaling, catalytic, immune response, storage and transport of information and energy. Catalytic proteins are called enzymes. They speed up biochemical processes occurring within the cell. The mechanism of catalytic activity of enzymes is by reduction of activation energy. As any catalyst they remain unaltered by the completed reaction. Antibodies are proteins responsible for recognition and for disabling foreign bodies like bacteria.

3.3.1.1 Conformation and Stability of Entrapped Protein in Porous Matrices

Zhou and Dill (2001) modeled the effect of confinement on protein configuration within inert nanopores using polymer diffusion theory and statistical thermodynamics. Only equations relevant to spherical pore confinement will be elicited here. In depth analysis can be found in the references mentioned. The assumption here is that the unfolded protein chain can be treated like a random Gaussian chain. If the protein chain has N residues (different amino acid groups/chains) and the effective bond length is b, then the probability density that the chain will begin at R_0 and end at R after n residues is given by G(R, R_0, n) for N≫1, n is treated as a continuous variable rather than a discreet one and it substitutes for time in diffusion equation whose probability density satisfies (Equation 3.5). Boundary conditions imply that G(R, R_0, n) = 0 at the walls of the pore

$$\frac{\partial G(R, R_0, n)}{\partial n} = \frac{b^2}{6}\Delta^2 G(R, R_0, n) \tag{3.5}$$

where b is effective bond length of the protein molecule

The partition function (Z_U) for every ensemble of chain with all the residues confined within the pore can be given as,

$$Z_U = \int dR \, dR_0 \, G(R, R_0, N) \tag{3.6}$$

where N is the number of residues in the protein chain

With integration over the pore volume, for a spherical pore of radius d

$$Z_U = \frac{d^3}{\pi} \sum \frac{1}{k^2} \exp(-\pi k^2/\beta d^2); k = 1, 2, 3.... \tag{3.7}$$

where $\beta = 3/2Nb^2$

If the folded protein is treated as a sphere with volume a_N then the partition function for native protein is the effective volume of the protein; (Effective volume = volume accessible to center of protein/total volume)

$$V_N = (d - 2a_N)^3/6 \tag{3.8}$$

where d = pore radius.

Partition function and Gibb's free are related by

$$\Delta G = -k_B T \ln (Z) \tag{3.9}$$

where Z is the partition function

Hence it can be deduced that, entrapment within the pore changes the free energy between the folded (native) and unfolded state of the chain by

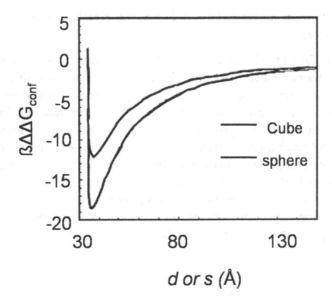

Figure 3.4. Effects of confinement on the free energy of protein folding. The dimensions of the folded and unfolded proteins are radius of gyration of unfolded chain $R_g=30A°$, ($R_g=Nb^2/2$) and radius of folded protein configuration 17.2 A°. The plot shows variation in free energy change associated with protein folding as a function of the pore diameter in which it is confined or a protein. It can be seen that the folded configuration is favored as the pore size approaches protein size. [Reprinted from Zhou, 2004, with permission from John Wiley & Sons Ltd.]

$$\Delta\left(\Delta G\right) = -k_B T \ln\left(\frac{V_N}{Z_U}\right) \tag{3.10}$$

As d approaches a_N, V_N decreases rapidly than Z_U, hence $\Delta(\Delta G)$ exhibits a maximum when the pore size approaches the native protein size. Free energy change due to confinement in a spherical pore was shown to be $\sim-15k_B T$ for a protein with a radius of gyration of unfolded chain $R_g= 30$ Å, ($R_g=Nb^2/2$) and radius of folded protein configuration 17.2 Å as pore size approached protein size (Fig. 3.4). Confinement also increases the folding rate of proteins by possibly speeding up the contact formation between residues of protein by excluding more expanded conformations (Zhou, 2004). It can be said that the folding of proteins within nanopores is a result of the decrease in the free energy of a protein molecule upon encapsulation. Ping's group has used a simplistic HP model (the reader is referred to Chan and Dill's paper for details)[1] to study the thermal stability of confined proteins. The HP model uses a chain of beads moving on a 2-D square lattice. These beads are assumed to be of two types, hydrophobic (H) and polar (P). The model mimics protein residues in water. Simulation studies of confinement of these chains within a box showed that as the box size decreases, the

---

[1] Chan HS, Dill KA, J. Chem. Phys. 100, 92381994.

melting point of the confined chain increases, indicating that more energy input is required to destroy the native structure of chain upon encapsulation.

Confinement of proteins within rigid spherical pores is analogous to confinement of proteins within a porous ceramic structure. These proteins confined within the nano pores of ceramic will experience electrostatic forces due to the charged surface groups on ceramics and proteins. Depending upon the isoelectric point of the protein and the ceramic, these interactions can be attractive or repulsive at a given pH. Cheung and Thirumalai (2006) studied the effects of strength nanopore-protein interaction (λ) and temperature (T) on the stability of the protein upon confinement. When there is no interaction between the pore and the protein ($\lambda=0$ meaning the nanopore is inert), then the effect of pure physical entrapment is to promote stabilization of the native state. The dependence of the folding rate of protein upon the strength of the pore-protein interaction was shown to vary in a non-monotonic way. Short-lived interactions between the pore wall and the protein side chain destabilize the native configuration but are not strong enough to permanently destabilize the protein. Under this condition the folding rate reaches maximum. As the attractive interaction strength increases more and more side chains become completely immobilized and the protein folding rate steadily decreases Fig. 3.5. Various other studies on confinement of proteins within nanopores also indicate stabilization of the (native structure of) protein (Zhou, 2004; Rathore, 2006).

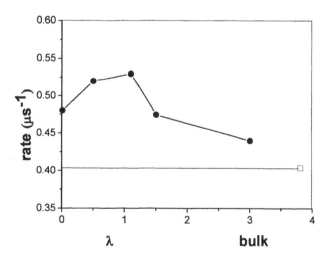

Figure 3.5. Folding rates of confined protein chains as a function of the strength of pore–protein interactions (λ). The maximum folding rate occurs at $\lambda =1$. As λ increases, pore–protein interactions changes form an inert to short-lived, transient interaction to a strong interaction. Short-lived interactions destabilize the native configuration but do not permanently destabilize the protein. Under this condition the folding rate reaches the maximum. With a further increase in λ, more and more side chains become completely immobilized and the protein folding rate steadily decreases. [Reprinted from Cheung and Thirumalai, 2005, with permission from Elsevier.]

Protein encapsulation within rigid ceramic cages of similar dimensions also offers protection against other denaturing forces, like the presence of an organic solvent or extreme pH. A confined protein loses its ability to unfold in response to denaturing forces. As the folded configuration becomes energetically favorable due to confinement, it requires extra energy inputs to undergo denaturation. Avnir and Mullerad (2005) have given another possible explanation for increased pH stability. For a sol-gel-encapsulated protein, the layer of water which hydrates the protein molecule within the nanocage is extremely thin, may be only 1–2 water molecule layers. If the water surrounding protein within a nanocage contains 100 water molecules, and the external pH is very high, pH=0, hydronium ions penetrate the pore so that nominal pH=0 is obtained inside the pore. Only two protons are sufficient to achieve pH=0 inside the pore. Protein can handle 2 protons in its vicinity, without undergoing denaturation. On the other hand in solution pH 0 means the protein is exposed to large number of protons leading to its denaturation. pH is a classical thermodynamics concept which might not be able to explain the observed phenomenon at such a small scale inside the nanocages (see Fig. 3.6).

3.3.1.2 *Dynamics of Entrapped Proteins Within Porous Matrices*

Once entrapped within the nanopores of ceramics, the dynamics of molecule such as rotational mobility or diffusional ability will be altered as a result of steric hindrance. If the pores are filled with a solvent, then microviscosity within the encapsulation site will be the additional constraint on the dynamics of the molecule within the pore. The solvent residing within the nano pores is likely to become perturbed due to interactions with the pore walls. The solvent confined

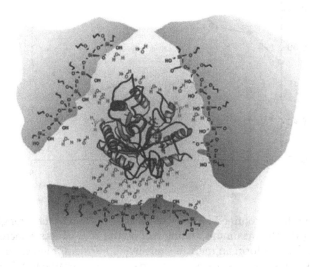

Figure 3.6. Schematic view of the entrapped enzyme with a few water molecules inside a pore, two of which are protonated; the nominal pH is very low. [Reprinted from Frenkel-Mullerad and Avnir, 2005, with permission from American Chemical Society.] For color reference, turn to page 150.

within the nano pores might behave differently than bulk solvent (Bellissent-Funel, 2003). Polar solvents interact with polar surface groups of a sol-gel silica matrix resulting in increased viscosity within the layer of the solvent adjacent to the pore walls. The depth of this perturbed layer and increase in viscosity depend upon the strength of interaction and dimension of the pore that confines the solvent. Such surface interaction effects on the viscosity of confined solvents are inversely proportional to the pore radius. As the pore radius decreases, the surface area to volume ratio increases, leading to increased surface interactions.

Non-polar solvents on the other hand due to lack of any interaction with the surrounding silica matrix remain relatively unperturbed within the pores (Klafter and Drake, 1989). It has been proposed that the solvent viscosity affects the entering and binding rates of the ligand to the protein (Beece *et al.*, 1980). It also affects the rate of conformational changes of proteins (Ansari *et al.*, 1992). Dunn and Zink, 1997, identified four different locations within the interior of a sol-gel derived porous silica matrix as: (1) the solvent inside the pore, (2) pore walls, (3) the boundary between the solvent and pore walls, and (4) the hannel which connects two pores as shown in Fig. 3.7. Proteins are generally added at the sol stage hence they remain dissolved in the liquid phase of the gel. Pore connecting channel is inaccessible to protein molecules. Depending upon the strength of the interaction of the biomolecule-solvent and biomolecule-encapsulating matrix, the molecule will be divided into the remaining regions. The dynamics of protein molecule will depend on its distribution within these regions. A molecule

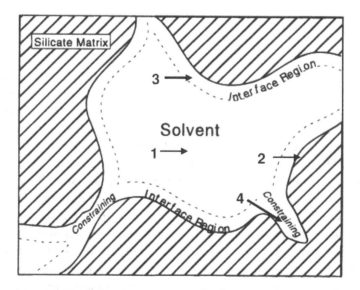

Figure 3.7. Sketch of four important regions of sol-gel materials. (1) Solvent inside the pore, (2) pore walls, (3) boundary between solvent and pore walls, and (4) the channel which connects two pores is a constraining region where the distance between the pore walls is about the same order of a probe molecule. [Reprinted from Dunn and Zink, 1997, with permission from American Chemical Society.]

completely existing within the solvent will behave similarly to the molecule in a free solvent. A molecule residing in a channel or close to the pore wall will have restricted freedom. Molecular dynamics simulation of a model protein within a solvent of high and low viscosity showed no dependence of protein equilibrium properties on solvent viscosity. The secondary structure of protein was the same in both viscosity domains. However the dynamic properties of protein were dependent on solvent viscosity. High viscosity affected the motion of protein atoms on the surface as well as in the center of the protein (Walser and Gusteren, 2001). Solvent dielectric constant also affects protein motion; protein dynamics is faster when the solvent has a high dielectric constant (Affleck *et al.*, 1992).

3.3.2 *Experimental Studies of Effect of Encapsulation of Biologicals within Porous Matrices*

3.3.2.1 *Protein Entrapment within Sol-Gel Synthesized Silica Matrix*

Most of the experimental studies of protein entrapment are performed on sol-gel silica-encapsulated proteins. The synthesis window for a silica sol-gel is wide; hence the properties of silica gel can be altered over a wide range by changing the processing parameters. Silica matrix being transparent is compatible with most analytical techniques used to study proteins in solutions. The biocompatibility of silica also makes it an attractive candidate for protein entrapment. Experimental studies of entrapped proteins are numerous. Various properties of entrapped proteins like thermal and chemical stability, dynamics, and reaction kinetics, have been studied. Protein accessibility to the analyte, diffusion of the analyte through the sol-gel matrix, and the effect of aging and drying of the silica matrix provide a necessary insight into the complex relationship between encapsulated proteins and ceramic matrix. Commonly used techniques for studying the properties of encapsulated protein within a silica matrix include: time-resolved fluorescence spectroscopy, Raman spectroscopy in ultraviolet (UV), and infra-red (IR) regions, circular dichroism (CD) spectroscopy, and electron plasmon spectroscopy (EPR).

For example CD can be used to study changes in the structure of protein due to encapsulation, upon changes in temperature, presence of alcohol, denaturing agents or ionic solvents. A CD far UV signal arises from the secondary structure of protein and a CD near UV signal arises from asymmetry in the tertiary structure (Eggers and Valentin, 2001). Intrinsic tryptophan (Trp) fluorescence is sensitive to protein unfolding in solution, so it is used to study the stability and flexibility of protein upon entrapment (Zheng and Brennan, 1998). Electronic absorption spectra of protein can yield important information as well. The characteristic soret band associated with the native heme protein conformation shift in band position reveals structural changes in heme proteins (Ray *et al.*, 2005). The accessibility of protein is studied by quenching of fluorescence. Protein can be labeled with a fluorescent molecule. Neutral and charged quenchers can be used to study the accessibility and analyte- protein reaction kinetics Zheng *et al.*, 1997).

3.3.2.2 *Thermal and Chemical Stability of Encapsulated Proteins*

Increased temperature leads to an increase in the thermal motions of a protein molecule. This disrupts the covalent bonds holding the protein structure together, leading to its denaturation. Similarly, changes in pH affect the charges carried by different amino acid residues leading to protein unfolding, which results in the loss of native structure and functionality of protein. Organic solvents may lead to aggregation and/or denaturation of proteins. It is interesting to report how encapsulated proteins respond to changes in temperature, pH, and solvent as opposed to their free counterparts. The thermal and chemical stability of encapsulated protein is determined by various factors such as host matrix, synthesis condition, pore size in which proteins are confined, size of protein, nature and strength of interaction between protein and the pores, etc. Various experimental studies indicate an increase in the thermal and chemical stability of biologicals after encapsulation within the silica matrix (Lan *et al.*, 1999; 2000). Spectroscopic studies of acid-induced unfolding of carbonmonoxymyoglobin (COMb) suggested that encapsulated COMb unfolding due to acidic pH occurs at a very slow rate as compared with free COMb. Upon unfolding some loosening of the helical structure of the protein may occur but overall the shape of the molecule is preserved (Samuni *et al.*, 2000). Ferricytochrom encapsulated in silica nanoparticles remained folded in the presence of the denaturing agent guanidinium hydrochloride (GuHCL), where as ferricytochrom in solution was fully denatured at the same concentration of GuHCL. Circular Dichroism (CD) spectroscopy of encapsulated ferricytochrom in the UV region at a high (30%) GuHCL concentration revealed little loosening of the α-helical structure. Still the encapsulated protein structure remained largely unaltered even at such high concentration of GuHCL (Fiandaca *et al.*, 2004).

The pore size determines the accessible volume for the protein for sampling various conformations. It also determines the surface area and thus the strength of interactions between the encapsulated protein and the matrix. Encapsulation of Ribonuclease A within mesoporous silica (MCM-48) (pore size 25 Å) resulted in increased thermal stability of the protein. Differential Scanning Calorimetry (DSC) peaks for entrapped protein appeared at $T_m = 90°C$ whereas for adsorbed and free protein the values were $T_m = 52°C$ and $T_m = 62°C$, respectively. Increased half width (20°C) of DSC peak for entrapped protein was an indication of pore size variation in the silica matrix (20–30 A°). As within each pore the enzyme was subjected to different set of conditions depending upon the pore size. Pressure perturbation (PPC) data showed that the protein within the pores is highly hydrated. This strong hydration of the protein molecule is thought to be contributing towards increased thermal stability along with conformational restriction upon encapsulation (Ravindra *et al.*, 2004). The nature of the precursor used for encapsulationencapsulation can also alter the thermal stability as seen in the encapsulation of human serum albumin (HSA). Thermal stability upon encapsulation for HSA was further improved when a biocompatible precursor Diglyceryl-silinase (DGS) was used rather than TEOS (Sui *et al.*, 2005).

Electrostatic interactions between protein and pore wall can influence the stability of an entrapped protein to a great extent. Ferric horse heart cytochrome C when encapsulated in wet silica gel preserved its native structure even under high acidic pH conditions (Droghetti and Smulevich, 2005). Differences in the unfolding spectra of encapsulated and free (in solution) protein indicated strong interactions between the encapsulated protein and surrounding silica matrix. The isoelectric point of this particular protein is at pH 10.2 making it positively charged below pH 10.2. Thus attractive electrostatic interactions between the protein and pore walls partially stabilized the protein at high pH. It is as if the attractive forces between the pore walls and the protein fixed the protein against any further unfolding at high pH. This stabilizing effect diminished below the isoelectric point of silica, when both silica and protein were positively charged. However these attractive interaction forces had an opposite effect in the presence of the denaturing agent GuHCL, where they contributed towards decreased stability.

Encapsulation within nanopores doesn't always lead to stability of encapsulated protein. Entrapment of apomyoglobin within a silica matrix resulted in its denaturation (Eggers and Valentine, 2001). This unfolding of protein after entrapment was not a result of electrostatic interactions between positively charged protein and negatively charged silica oxide groups [pH=7], as it was shown that changes in pH and ion concentration did not affect the unfolding of apomayoglobin. Thermodynamically unfavorable water-silica interactions lead to the unfolding of protein (see Fig. 3.8). Encapsulation of bovine serum albumin (BSA) restricted conformation changes that accompany unfolding of protein at acidic pH. It was largely because of steric hindrance imposed by pore walls as the pore size and the protein size were similar. However, encapsulation of myoglobin at pH 7 resulted in 70% loss of the native structure. Myglobin's thermodynamic stability is inferior as compared with more structured proteins like ribonuclease A. Encapsulation within rigid cages of silica seems to perturb such "soft" proteins to a greater extent. This fact underlines the importance of the intrinsic properties of protein in controlling the properties of encapsulated protein (Edmiston *et al.*, 1994). Entrapment of Carbonic anhydrase did not increase its thermal stability (Badjic and Kostic, 1999). The intrinsic property of the protein to undergo a sort of irreversible unfolding upon increase in temperature even in solution seems to have affected its unfolding property in the encapsulated state as well.

Various factors which affect the stability of encapsulated protein such as the precursor and the method used to synthesize bio-doped ceramic, pore size and its distribution, intrinsic properties of encapsulated protein, and interaction between protein and encapsulation matrix are often interrelated. The precursor used determines the properties of the encapsulation matrix such as its isoelectric point, pore size distribution, shrinkage behavior etc. Protein's intrinsic properties such as flexibility of the polypeptide backbone of the protein structure, molecular weight, size, isoelectric point etc. will decide its response to confinement and interaction with the confining matrix. These parameters are complex, which makes the outcome of encapsulation on protein stability unique for each protein for a given set of experimental conditions.

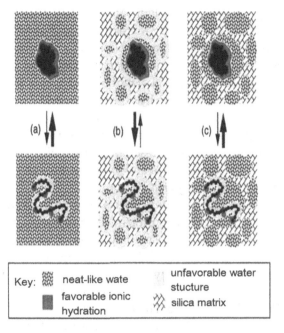

Figure 3.8. Role of bulk water structure in protein folding equilibria and the hydrophobic effect. A hypothetical two-state equilibrium is depicted for a folded globular protein (top panels) and the unfolded random-coil conformation (lower panel) in three different aqueous environments. (a) In an ideal dilute solution, the folded conformation is favored due to a strong hydrophobic effect. (b) The glass-entrapped protein is influenced by the unfavorable water structure at the silica interface of the surrounding pores. The unfolded state predominates due to a diminished hydrophobic effect. (c) Addition of compatible solutes reduces the average free energy of the bulk water to a value that more closely resembles neat water. The native state is favored due to the restored driving force of the hydrophobic effect. [Reprinted from Eggers and Valentine, 2001, with permission from Elsevier.] For color reference, turn to page 150.

3.3.2.3 *Protein Dynamics*

Studies of small solute molecules entrapped within TEOS-derived silica sol-gel have shown that electrostatic interactions between these molecules and the silica matrix affect the rotational mobility of the entrapped molecules. The electrostatic attraction between the cationic solute molecule and the silica wall restricts rotational mobility to a greater extent. X-band Electron Plasmon Resonance (EPR) spectrum of entrapped cationic solute molecules showed that a fraction of it was completely immobilized due to attractive electrostatic forces (Wheeler *et al.*, 2004). Although these studies were conducted on solute molecules, observations may very well be applied to positively charged proteins and negatively charged silica walls at pH 7. On a global scale, the rotational mobility of protein may be hampered due to the high microviscosity of the solvent within the encapsulation site if the protein size and pore size are comparable (Gottfried, 1999). Tertiary and quaternary relaxations of hemoglobin encapsulated in silica were affected by solvent viscosity. A slow relaxation rate at a low temperature implied the dependence of solvent viscosity upon temperature and dependence of the relaxation rate upon solvent viscosity (Das *et al.*, 1999). Time-resolved fluorescence anisotropy of HSA entrapped in various matrices including TEOS-derived silica showed that entrapment restricted global motion and encouraged the local motion of HSA (Sui *et al.*, 2005). Properties such as the rigidity and the shape of the protein affect its dynamics upon encapsulation. Intrinsic nanosecond globular motions of anti-dansyl antibodies observed in solution were restricted upon encapsulation in silica sol-gel. The antibody was essentially immobile. Antibodies are typically rigid; a particular Y shape of this antibody was thought to be the reason for restricted globular motion (Doody *et al.*, 2000).

But strictly speaking, encapsulation effects on protein dynamics are case specific. There are various cases where protein dynamics was unaltered or altered to a very small extent due to encapsulation within a silica matrix. Entrapped antibodies were not completely immobile after encapsulation (Wang *et al.*, 1993). Circular Dichroism (CD) spectra analysis of silica-encapsulated bacteriorhodospin showed that neither the local nor the global motion of protein was restricted by the sol-gel matrix (Wu *et al.*, 1993). Dipolar relaxation measurement of cytochrome c encapsulated in silica thin films showed that interactions between the protein and matrix restrict rotational motion but only to a small extent (Dave *et al.*, 1995). Time-resolved fluroscence spectroscopy of acrylodan-labeled BSA and HSA showed that encapsulation does not restrict the global motion of protein (Jordan *et al.*, 1995). The dynamic properties of encapsulated proteins depend on solvent's viscosity within the pore of encapsulation matrix.

The solvent behavior also changes upon encapsulation due to the interaction between the encapsulating matrix and the solvent. These interactions depend on various factors such as pore size, functional groups on walls of pores of the encapsulating matrix, the intrinsic properties of the solvent such as polar or nonpolar nature, bulk viscosity, etc. The dynamic properties of encapsulated proteins are controlled to a large extent by the relative strengths of the solvent-pore and solvent-protein interaction. All these parameters affect the properties of encapsulated proteins in many ways; for example, increased solvent viscosity within the pores of ceramic matrix may also lead to increased thermal and chemical stability as unfolding dynamics associated with an increase in temperature or a harsh chemical environment will be slow. The intrinsic properties of protein also influence the dynamics of encapsulated protein. Large proteins will be more affected by increased viscosity of the solvent as they exhibit more contact with the solvent. Understanding the properties of the ceramic matrix as well as those of the protein and the solvent will help to achieve a better understanding of the effects of encapsulation upon protein dynamics.

3.3.2.4 *Effects of Aging and Drying*

When discussing the effects of entrapment within sol-gel-derived silica on proteins, it is very important to consider the effects of aging and drying. The gel structure continues to evolve beyond the gelation point as more and more bonds are forming and cross-linking increases (aging of gel). Aging can be of two types: wet aging where the gel is kept immersed in buffer solution for the required period of time; or it can be dry aging, where the solvent within the pores evaporate under ambient conditions. The preparation parameters and aging conditions of the gel determine the internal structure evolution of the gel (Flora *et al.*, 1999). Aging and drying have a profound effect on the pore structure as the pore radius decreases; the solvent is expelled, so the internal viscosity of the solvent and electrostatics within the pore thus changes. Also, drying stresses are generated within the drying gel (Brinker and Scherer, 1990). All these factors significantly affect/control the structure, functionality, and accessibility of the entrapped proteins.

Entrapment of monelline within wet-aged silica monolith did not improve its chemical or thermal stability but entrapment within dry-aged monolith showed improved thermal stability and further improved long-term stability than in the wet gel. The unfolding temperature of monelline changed from 50°C to 53°C upon encapsulation in wet-aged silica monoliths, and to 64°C in dry-aged silica monoliths. Increased microviscosity and more "structured" solvent through hydrogen bonding with polyhydroxylated silane were thought to be responsible for the increased thermal and chemical stability of monelline in dry-aged monoliths (Zheng and Brennan, 1998). Increased microviscosity and solvent-pore interaction will slow down any protein movement including unfolding reaction due to increased temperature that explains the increased thermal stability of monelline in dry-aged gel.

Bovine serum albumin (BSA) immobilized within hydrated TEOS-derived silica showed 75% accessibility whereas in the dried gel, the accessibility reduced to 50% (Wambolt and Saavedra, 1996). This reduction in accessibility can be attributed to the shrinkage of gel upon drying, which caused diffusion difficulties to the fluroscence quencher iodide. Silica gel-encapsulated antifluroscein antibody lost its affinity towards fluoroscein when the gel was dried for 3 weeks. Drying of silica gel likely caused conformation of encapsulated antibody to undergo drastic changes, making the binding site inaccessible to fluroscein. This loss of affinity was irreversible; rehydration of dried gel did not recover lost affinity. When antifluroscein was encapsulated in wet-aged gel (kept immersed in water), its affinity toward fluroscein was found to be intact. Hydration or gel pores that are filled with a solvent prevented conformational changes, leading to affinity loss. Wet aging also alters pore size, protein-pore interaction but here in this case the presence of the solvent played a major part in protecting the conformation of the antibody. Drying reduced protein stability in the case of protein parvalbumin (Flora and Brennan, 1998).

Drying altered kinetics of relaxation in the case of sol-gel-encapsulated bacteriorhodospin (bR). Upon absorption of photon, bR transports proton form the exterior to the interior of the cell, the resulting proton gradient within the cell is used to drive the synthesis of ATP. During photon-induced photocycle, bR relaxes through various intermediate states. The kinetics of relaxation of one such state, the intermediate state M in silica xerogel, was slowed by a factor of 100 upon drying (Shamansky *et al.*, 2002). The effect of different drying conditions of CO_2 super critical drying and drying by evaporation on the activity of lipase encapsulated in silica was studied. It was observed that enzyme stability and activity can be improved by avoiding capillary stresses arising in evaporation drying, by using CO_2 supercritical drying (Buisson *et al.*, 2001).

It is not always drying stresses that cause the denaturation of protein. Loss of the solvent also plays a major part in stability of the stability of the encapsulated protein. The effects of aging and drying on silica-encapsulated bR were studied by observing absorption band of chromophore bound to br apoprotien. Heat and denaturation led to loss of chromophore and loss of light-transducing activity of bR. It was seen that aging (7 days) as well as drying (21 days) did not affect the structure of the protein. However, bR started bleaching when sol-gel film lost more than 80% of solvent. This indicates that though drying stresses did not affect the structure of the protein, there is a critical solvent level that is necessary for the protein to maintain its structure (Wu *et al.*, 1993). The intrinsic properties of proteins such as their size, shape, and their interactions with the surrounding matrix also decide the extent of drying effects on their stability. Spectroscopic studies of cytochrome c within dried silica gel showed no effect of drying stresses on the protein structure of stability. The characteristic soret band showed the protein reversed back to a configuration similar to that in aged gel after dehydration of xerogel. Cytochrome c is a high molecular protein carrying a net positive charge. It acted as a template for silica condensation by virtue of the electrostatic attraction between them. It also prevented collapse of pores during the drying of gel (Lan *et al.*, 1999). The Michaelis constant (K_m) of protein alkaline phosphate encapsulated within xerogel showed a distribution. It indicates different distribution of activity for an individual protein molecule (Sánchez *et al.*, 2003). Evolution of each pore upon drying is unique and so the effects of drying on each protein molecule in different pore are unique too.

3.3.2.5 Accessibility and Reaction Kinetics of Entrapped Proteins

Encapsulation of proteins/enzymes into a nanoporous matrix results in decreased response time for reactions between the entrapped enzyme and the analyte molecule. The reaction kinetics of proteins entrapped in nanosized pores will depend on the conformational flexibility of protein as well as the diffusion rate of analyte within the sol-gel matrix. Increased viscosity of entrapped solvents leads to a slower diffusion of analytes within the glass and lowers the protein conformational flexibility. Electrostatic or other interactions between the analyte and the matrix also contribute to slower diffusion rates of analyte. The silica isoelectric point is nearly 2 (Iler, 1979), and the silica pore surface contains various groups like silanol (Si-OH), siloxide (Si-O-), and Siloxane (Si-O-Si). The silica surface will be negatively charged under most conditions.

If the analyte and silica charges are similar, uptake of analyte by the matrix will be partial. If the charges are opposite because of strong absorption on the surface, the oppositely charged analyte may not be able to penetrate the matrix fully. On the other hand, an anionic polar analyte might be completely rejected by the silica matrix (Shen C, Kostic, 1997; Badjic and Kostic, 2001). Hydrogen bonding between organic molecules and silica can cause excessive uptake from the solvent. Badjic and Kostic, 2001, have shown that excessive aniline uptake from CCl_4 into a silica matrix and balanced uptake of styrene from CCl_4 into a silica matrix are consequence of the capability of aniline to donate as well as accept hydrogen and the incapability of styrene to do so. Kinetics of reduction by sodium dithionite of silica nanoparticle-encapsulated ferricytochrome was reported to be about 10 times slower than the protein in free solution; the slow diffusion of dithionite through silica nanoparticles caused this decreased rate of reaction (Fiandaca *et al.*, 2004).

The diffusion rates of the analyte molecules through a silica matrix can be easily controlled by controlling the pore size of the matrix, and the electrostatic interaction between them. Cationic molecules can diffuse at rates similar to those in solution by incorporating positively charged entities into the silica matrix (Kanungo and Collison 2005). The reaction rates for oxidation of dye ABTS, which has positive charge, were higher for methylated silica glass ($MeSiO_2$) than pure silica or Propylated silica glass SiO_2. Silica surface has a negative charge whereas HRP has a positive charge. The diminished negative charge after methylation of silica increases the reaction rate of catalysis (Kadnikova and Kostic, 2002). Encapsulated protein accessibility to a particular analyte also plays a major role in deciding reaction kinetics. All protein sites within an encapsulation matrix are not necessarily accessible to analytes. Protein accessibility may be evaluated by quenching of the entrapped protein fluorescence by reagents, such as O_2, acrylamide, Co^{2+}, or I^- (Keeling-Tucker and Brennan, 2001).

Enzyme kinetics alters upon encapsulation; K_m and K_{cat} of encapsulated enzymes are often different from those of free enzymes. K_m is the substrate concentration at which the rate of enzymatic reaction is half of its maximum. A low value of K_m indicates a strong affinity between substrate and enzyme. A high value of K_m indicates a week affinity between the substrate and the enzyme. K_{cat} is called the turnover number and it represents the maximum number of substrate molecules that can be converted into products per molecule enzyme per unit (Palmer, 1995). Reportedly, Kmslightly increased upon encapsulation and Kcat decreased, indicating weak substrate binding and slow reaction kinetics (Pioselli *et al.*, 2005). The increase in Km can be due to denaturation of some enzyme upon encapsulation or its inaccessibility to the analyte. A lower enymatic reaction rate upon encapsulation could also be an indication of diffusion problems for the analyte. The catalytic activity experienced a 6- to 8-fold decrease than in free solution, which was attributed to the restricted conformational changes which are necessary for catalysis (Pioselli *et al.*, 2004). It should be noted that the ability of the sol-gels to limit conformational change depends on the preparative protocol (Khan *et al.*, 2000). Reaction kinetics cytochrome of quenching of zinc was affected by the electrostatic interaction between zinc cytochrome and silica. Anionic analyte was excluded from the matrix. Ionic strength or pH can be changed to increase the uptake of anionic analyte into the silica matrix (Shen and Kostic, 1997).

Encapsulation does not always lead to partial loss of accessibility or binding abilities of protein. Fluorescence studies showed encapsulated parvalbumin to be completely accessible (Flora and Brennan, 1998). In another case, an antibody-doped sol-gel sample stored at room temperature showed no change in the binding capacity of the antibody (Bronstein *et al.*, 1997). Accessibility and reaction kinetics of proteins upon encapsulation cannot be generalized, as in the case of protein dynamics and stability. Interactions between the analyte and the protein-encapsulating ceramic matrix control the accessibility of protein. Continued evolution of the gel after the gelation point is reached should be kept in mind when considering accessibility of protein. Continued evolution of the gel will result in smaller pores and loss of solvent; both of these factors will affect the diffusion of analyte through the sol-gel matrix. Also drying of gel will alter the accessibility of protein because of similar reasons, but here the solvent may be completely absent depending upon the degree of drying. Supercritical drying will not change pore size as in the case of ambient drying. Choosing the appropriate processing parameters to obtain pore sizes necessary to promote analyte diffusion and modifying the encapsulation matrix to achieve favorable electrostatic interaction between the analyte and the encapsulation matrix will help achieve maximum protein accessibility following encapsulation.

3.3.3 Cells and Bacteria

Encapsulation of cells and bacteria has gained much importance. The application spectrum for immobilized cells is large, spanning biosensing, toxic compounds' detection, wastewater treatment, adsorption of heavy metals, controlled release, and cell transplantation. Cell transplantation can be an alternative to organ transplantation. Encapsulation of living cells within porous membranes can shield the cells from immune attack (Sanchez *et al.*, 2001). Growth of cell culture within an encapsulation matrix can be exploited for the production of hormones, growth factors, and enzymes. Such efforts have been reported, where tissue culture growth is achieved within sol-gel thin films (Zolkov *et al.*, 2004). Also, encapsulation of cell or bacteria rather than enzymes will help avoid a tedious enzyme extraction process. Carturan *et al.*, recently reviewed encapsulation of functional cells within a silica sol-gel matrix, with a focus on the biosil method for cell encapsulation and biocompatibility of an encapsulation matrix for cell grafts (Carturan *et al.*, 2004).

Bacteria are unicellular organisms with a negatively charged cell surface at physiological pH. Bacteria can form spores to survive under tough conditions such as no water, no nutrient, extreme heat, and even radiation. When they found themselves in suitable conditions, they again come back to life by transforming into cells. Most encapsulation studies of bacteria within a ceramic matrix are within silica matrix.Various proteins on the cell surface of bacteria might interact with the silica matrix.The interaction between bacteria *staphylococcus aureus* and a silica surface were explored using atomic force microscopy. The results showed that bacteria formed attractive, non-specific bonds, like Van der Waal's forces, with the silica surface half of the time upon contact formation (Yongsunthon and Lower, 2006).

Hexagonal arrays formed by yeast cells on glass slides when these slides were dip coated with silica sol and yeast cell solution (1:1). This assembly of cells might be a result of shrinkage of silica film and forces acting during film pulling. Both silica and yeast cell have a negative charge at physiological pH; thus the interaction between them might have resulted in minimization of the area occupied by cells (Chia *et al.*, 2000). Encapsulation of bacteria can prevent their aggregation as opposed to bacteria in solution. But the stresses of encapsulation within a silica matrix can result in cell membrane lysis. The membrane permeability of bacterial cells can become modified as a result of encapsulation in a pure silica gel (Nassif *et al.*, 2003). SEM of sol-gel silica-encapsulated *Escherichia Coli* (*E. Coli*) cells showed a random distribution of the cell within the matrix. Within the wet matrix, their cell membranes were intact but dried gels showed partially destroyed membranes (Fennouh *et al.*, 1999). Cell membrane lysis during encapsulation may result in faster kinetics of enzymatic reaction as diffusion through membrane becomes easier.

The rigidity of the ceramic matrix and the small pore sizes of sol-gel-derived glasses, which restricts unfolding of proteins, have similar effects on bacteria. The rigidity and small pore size of silica gel constrain encapsulated cells, as a result, cell division and cell growth possibilities are excluded (Branyik and Kunkova, 1998). Tensile stresses of silica gel prevented cell reproduction within the gel (Inama, 1993). However, anaerobic sulfate-reducing bacteria encapsulated within the silica matrix were able to undergo cell division and grow. Activity measurements of these bacteria lowered after 10 weeks of storage, indicating lowered viability. After immersion in nutrient solution for 6 days, the activity increased significantly, indicating an increase in viability (Finnie, 2000). Fennouh *et al.*, 1999, studied the enzymatic activity of entrapped bacteria within wet and dry gels. Activity with dry and wet gel was measured and compared with an aqueous suspension of bacteria. A decrease in K_m upon encapsulation indicates improved activity in wet gels. Dried gel-entrapped bacterial enzymatic activity was found out to be only 15% of the original activity in aqueous solutions. This indicates that sol-gel entrapment prevented the aggregation of bacteria, leading to increased activity, where as in dried gel, silica shrinkage may be responsible for ruptured cell membrane and loss of activity.

The addition of a biocompatible component to the silica matrix has positive effects on encapsulated bacteria viability just as in the case of protein. Encapsulation of bacterium *E. Coli* within a silica sol-gel showed that the number of culturable bacteria upon encapsulation increased after addition of glycerol. 40% of the bacteria remained viable after 4 weeks when encapsulated in silica gel with glycerol as opposed to 10% in both silica gel without glycerol and a water-glycerol solution (Nassif *et al.*, 2002). Glycerol was thought to protect the cell against encapsulation-induced stress with its membrane-stabilizing property. Bacterial activity was found to last up to several weeks in wet silica gels kept at room temperature when glycerol was added to the silica matrix (Nassif *et al.*, 2003). Similarly, the use of an aqueous sol-gel route for encapsulation proved beneficial for bacterial activity. Nutrient supply is important in the survival of entrapped bacteria. Activity measurement of sol-gel silica-entrapped atrazin-degrading pseudomonas showed entrapped bacteria when incubated in nutrient growth medium showed 8 fold increase in activity against entrapped bacteria incubated in plain saline solution (Rietti-Shati *et al.*, 1996). In an interesting study of entrapped bacterial viability in the presence of quorum-sensing molecules, it was found that the percentage of culturable bacteria after 4 weeks in wet gel was much higher in the presence of quorum-sensing molecules than in their absence. Quorum-sensing molecules are related to gene expression as a function of the cellular density of bacteria. The presence of a QS molecule might have limited the death rate of bacteria (Nassif *et al.*, 2004).

During encapsulation, bacteria cells can outgrow on the surface and within the surface cavities of sol-gel silica, contributing to bacterial activity measurement. In order to measure activity of only encapsulated bacteria, controlling the outgrowth of bacteria on the gel surface and growth medium is very necessary. During encapsulation of fungus *penicillium chrysogenum* and bacterium *streptomysces rimosus* within silica gel, UV irradiation of the surface of silica gel and growth medium was shown as an effective sterilization method to prevent outgrowth of bacteria. This ensured that the measured activity is solely due to encapsulated bacteria (Taylor *et al.*, 2004). Cells of Methylomonas sp. strain GYJ13 encapsulated in sodium silicate sol-gel showed 1.5 times more activity than free cells. Beyond a critical cell loading, the catalytic activity was lowered, which might be due to diffusional limitations on substrate entering the sol-gel matrix or aggregation of cells at such high concentrations. Stability against pH and temperature also increased after encapsulation (Chen *et al.*, 2004). Encapsulated yeast cells showed better metal-binding capacity as compared with cells in solution (Al-Sarai, 1999). For another metal-binding bacteria *Bacillus sphaericus* JG-A12, the metal binding capacity of cells and stabilized surface layer protein were not changed by encapsulation, but for bacterial spores it decreased after immobilization in silica sol-gel (Raff *et al.*, 2003). Bacteria or cell encapsulation within the ceramic matrix thus bears resemblance to protein encapsulation. In this case, together with the properties of the encapsulating matrix, encapsulation route, individual traits of bacteria are a decisive factor in its viability upon encapsulation. Each bacterium responds and adapts to the surrounding environment in its own unique way; thus, encapsulation under the same conditions may bear different consequences for the different bacteria used.

3.4 APPLICATIONS OF BIO-DOPED CERAMICS

Applications of biodoped ceramics are numerous, including biosensors, immunoassays, biocatalysis control release agents, etc. The most widely explored application of biodoped ceramics is in the field of biosensing. Porous ceramic matrixes have high physical rigidity, inertness, high thermal stability, and they do not swell in solvents, which prevents leaching of entrapped biological entities. Silica sol-gel-based biosensors have been widely reported and have been thoroughly studied. Various other metal oxide matrices have also been used for biosensing applications. A phenol-detecting amperometric tyrosinase biosensor consisted of two sol-gel layer of Al_2O_3 deposited on a glassy carbon electrode. The inner-layer Al_2O_3 was encapsulating a mediator $Fe(CN)_6^{4-}$, while the outer-layer Al_2O_3 was encapsulating tyrosinase. The porosity of the Al_2O_3 sol-gel film was adjusted such that $Fe(CN)_6^{4-}$ resided in smaller pores and tyrosinase resided in larger pores. The sensor response to phenol was linear in the detection range $1.0 \times 10^{-7} - 2.5 \times 10^{-4}$ mol/L (Liu *et al.*, 2000). A mediator-free tyrosinase biosensor based on entrapment of tyrosinase wihin an Al_2O_3 matrix showed a high sensitivity ($127 \ \mu A/mM^{-1}$), with a response time of less than 4 seconds. The linear range of detection of phenol was $1.5 \times 10^{-9} - 3.5 \times 10^{-4}$ mol/L. It was found that hydrophilic Al_2O_3 is more beneficial for immobilization of tyrosinase.

Zirconium dioxide (ZrO_2) has also been used for biosensor application. A H_2O_2-detecting amperometric HRP biosensor was constructed by entrapping HRP within ZrO_2. The base electrode used was a pyrolytic graphite electrode (PGE). Thionine was electropolymerized on a PGE electrode. An HRP- ZrO_2Sol-gel layer was then deposited on top of it. A ZrO_2 matrix stabilized and retained activity of HRP. The sensor response was linear in the range of $2.5 \times 10^{-7} - 1.5 \times 10^{-4}$ mol/L; the sensor was stable for 3 months. The response time of such a prepared sensor was less than 10 sec (Liu *et al.*, 2003). Choi *et al.*, 2005, reported a glucose oxidase-doped titania-nafion as a superior matrix for an amperometric glucose biosensor than a glucose oxidase-doped silica-nafion composite. It showed a better response time, storage stability, and sensitivity. Nafion prevented film cracking. Chen *et al.*, 2001, fabricated a tyrosinase biosensor by entrapping tyrosinase within a sol-gel-derived composite of titania and a grafting copolymer of poly(vinyl alcohol) with 4-vinylpyridine.The response time of the sensor was 20 sec, slower than the tyrosinase sensor prepared by the vapor-deposition method. The sensitivity was reported to be $145.5 \ mA/mmol^{-1}$. This was higher than the tyosinase sensor prepared by the vapor-deposition method (Chen *et al.*, 2001).

Composites of titania and grafting copolymer of poly(vinyl alcohol) with 4-vinylpyridine encapsulating glucose oxidase have been used for glucose biosensing. This amperometric glucose biosensor showed fast response, high sensitivity, and good storage stability. Another study of a tyrosinase biosensor used novel Titanium dioxide (TiO_2) hybrid composite. Hydrolysis of titanium alkoxides was carried out in the presence of acetylacetone. The product was used as a precursor for the synthesis of a TiO_2 sol-gel as shown in Fig. 3.9. Tyrosinase was entrapped within this sol. This novel composite help avoiding cracking of a sol-gel film upon drying as well as an extended pH range in which biodoped titania can be fabricated. This amperometric metric biosensor was used for detection of phenols. The response time of sensor was less than 10 sec, with a linear range of detection from

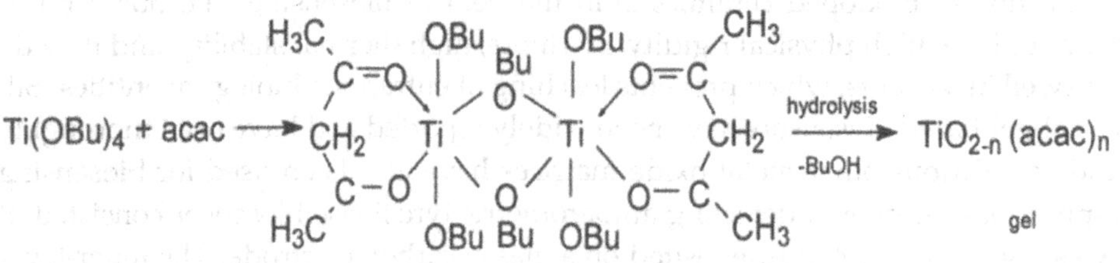

Figure 3.9. Scheme of hybrid titania sol-gel preparation. Ti(OBu)$_4$(3.4g) and acetylacetone (0.7g) were mixed and stirred for 2 hours. Then, Water of Acetic acid-phosphate buffer was added. (Molar ratio of Ti(OBu)$_4$:acac:H$_2$O is 1:0.7:200). After heating the mixture at 80–90°C, a stable, homogeneous titania sol was obtained. [Reprinted from Zhang *et al.*, 2005, with permission from Elsevier.]

$7.5 \times 10^{-8} - 6 \times 10^{-6}$ M of the phenol concentration (Zhang *et al.*, 2003). Apart from the conventional sol-gel based biodoped metal oxides, a vapor deposition method has been explored for the synthesis of biodoped ceramics for biosensing applications. An amperometric tyrosinase biosensor for phenol detection based on encapsulation of tyrosinase within TiO$_2$ using the vapor deposition method showed a linear response in the range of $1.2 \times 10^{-7} - 2.6 \times 10^{-4}$ mol/L and a response time of less than 5 sec. The sensor exhibited good storage stability. The mild nature and the absence of any additives during the vapor deposition method was said to help maintain enzyme activity (Yu *et al.*, 2003).

Hemoglobin encapsulated within titania sol-gel by the vapor deposition method retained its native structure and activity. An amperometric biosensor for hydrogen peroxide (H$_2$O$_2$) based on this biodoped titania matrix exhibited a fast response and it could detect H$_2$O$_2$ in the range of $5 \times 10^{-7} - 5.4 \times 10^{-4}$ M (Yu and Yu, 2003).

Biodoped zink oxide (ZnO) has also been used for biosensing. Tyrosinase was entrapped in a positively charged ZnO sol-gel matrix. The response time of the sensor was less than 5 sec, and the sensitivity was 168 μA (mmol/L)$^{-1}$. The linear range for detection of phenol was 1.5×10^{-7} to 4.0×10^{-5} mol/L. It was found that ZnO matrix being positively charged was more suitable for tyrosinase loading and for retaining its activity (Liu *et al.*, 2005). A glucose oxidase-doped vanadium pentoxide(V$_2$O$_5$) sol-gel was used for glucose biosensing (Glezer and Lev, 1993). Recently, the author's group (Gouma *et al.*, 2005) reported encapsulation of urease within a molybdenum trioxide (MoO$_3$) gel for application in resistive gas sensing. The enzyme clusters seemed entrapped in the pores of the gel in the transmission electron microscope (see Fig. 3.10). Urease catalyzes the hydrolysis of urea, producing carbon dioxide and ammonia. The encapsulation matrix, which is molybdenum trioxide, is highly selective and sensitive to ammonia gas. The electrical resistance of the oxide matrix decreases when exposed to ammonia even at very low concentrations. The typical response time is a few milliseconds. This work is a platform for development of resistive biosensors of high specificity by employing biodoped ceramics (Gouma *et al.*, 2005). Resistive sensors are expected to be

Figure 3.10.   Encapsulation of urease in the pores of a Molybdenum trioxide sol-gel.

more sensitive and their response times and recovery times are rapid, compared with other types of biosensors such as optical, amperometric, potentiometric, etc. Resistive sensors are also economic and easy to fabricate.

Various non-enzymatic proteins have been encapsulated and used as a biorecognition element. Aequorin a bioluminescent protein found in the jelly-fish *Aequorea sp.*, has been immobilized in a sol-gel-drived SiO_2 matrix. The luminescence from this protein is sensitive to the presence of calcium ions in the solution. Immobilization did not affect protein's calcium ion-binding capacity. An optical biosensor for detection of calcium ion was successfully built (Blythe *et al.*, 1996). Sensors based on entrapped bacteria/cell also have been realized. An optical biosensor based on the entrapment of engineered luminous bacteria within a silica sol–gel matrix showed a broad-range response to various toxic compounds Premkumar *et al.*, 2002). Biodoped ceramics consisting of yeast cells immobilized into an Al_2O_3 matrix was used for construction of a Biological Oxygen Demand (BOD) sensor. Although heavy metal ion did not affect the activity of yeast cell, high salt concentration did affect respiration of yeast cells and hence the sensor performance (Chen *et al.*, 2002).

Entrapped enzymes within a ceramic matrix can be reused to catalyze specific reactions. Flow-through applications can be realized with sol-gel as encapsulation confers in resistance to leaching, improved stability against denaturing forces as well as good mechanical stability (Vera-Anila, 2004). A "mini–reactor" consisting of biosilica-entrapped butylcholinesterase was used for conversion of indophenyl acetate. Conversion continued over 500 mL of indophenyl without loss of activity (Luckarift *et al.*, 2004). Silica monolith encapsulating protease was used as a bioreactor for transesterification between glycidol and vinyl *n*-butyrate. Protease catalyses transfer of the alkoxy group between glycidol and vinyl *n*-butyrate. The bioreactor consisted of a microcapillary tube filled with silica monolith-encapsulating protease. The inner diameter of the microcapillary tube ranged from 0.1 to 2 mm. The flow rate of substrate solution through microcapillary was 0.0004–5.0 mL/min (Kawakami *et al.*, 2005).

Sakai-Kato *et al.*, 2002, developed an enzyme bioreactor using a capillary containing trpsin-encapsulating silica gel. Silica sol containing trpsin was directly poured into a fused silica capillary. The capillary was pretreated with methacryloxy-propyltrimethoxysilane (MPTMS). MPTMS covalently attaches the capillary wall to the silanol groups of the silicate matrix, preventing the gel from leaking out of the capillary. The activity of the encapsulated trypsin toward peptide [Tyr8]-bradykinin was found to increase 700 times than that of free trypsin. Also, the encapsulated trypsin showed increased stability during continuous use than that in free solution. Kato *et al.*, 2005, later fabricated a monolithic bioreactor, using a trypsin-encapsulating gel-coated porous silica monolith.

Bioreactors using proteins encapsulated in silica sol-gel were found to be useful for the digestion of small molecules like peptides, but proteins, due to their large size, were unableet al to diffuse through the nanopores silica gel. Use of a macroporous silica monolith as the supporting matrix for enzyme immobilization allowed diffusion of larger molecules like proteins. Also, large surface area resulted in an increase in the number of immobilized trypsin molecules. The sol-gel reaction was optimized so that trypsin-containing sol formed a thin layer on silica monolith. A microtiter plate was chosen as an analytical platform. A Trypsin-containing gel-coated silica monolith was fabricated to fit into a 96-well microtiter plate. This bioreactor successfully demonstrated digestion of protein molecules like casein. This bioreactor could be used for several days without loss of trypsin activity. Also, storage at 4°C for 2 months did not affect its stability. Using photopolymerized sol-gel (PSG) as a supporting matrix for enzyme immobilization. Kato *et al.*, 2004, reported a pepsin reactor. PSG was synthesized from MPTMS containing both methacrylate and alkoxysilane groups. A pepsin-containing silica sol was introduced into a PSG-containing capillary till pepsin sol soaked PSG monolith. The capillary was held at 4°C for 2 days to allow gelation of pepsin sol and coating of PSG with the pepsin-containing gel. Pepsin retained its activity upon encapsulation. It successfully digested proteins and peptides, by cleaving peptide at hydrophobic amino acid sites. The resulting mixture of peptide fragments was separated in the portion of the capillary where no PSG monolith exists. These peptide digests were directly analyzed using electrospray ionization mass spectrometry (ESI-MS). Insulin and lysozymes were digested by immobilized pepsin and provided reliable and reproducible mass spectrometric results on the sequence data.

Microchips along with silica capillaries and mirotiter plates have been used as an analytical platform for bioreactors. A bioreactor for digestion of cytochrome c and BSA was fabricated by immobilizing a trypsin-encapsulated silica gel within microchannels of a microchip. The microchip was made from poly (methyl methacrylate) (PMMA); its surface was modified with zeolite nanopartiacles. Sucessful digestion of cytochrome c and BSA was achieved by encapsulated trypsin, which maintained its activity for more than 1 month when stored at 4°C (Huang *et al.*, 2006). Sakai-Kato *et al.*, 2003, fabricated a bioreactor using a trypsin-encapsulating sol-gel on a plastic microchip. PMMA was used to fabricate the microchip. This bioreactor integrated digestion, separation, and detection of proteins. Electroosmotic flow (EOF) was used for separation of peptide fragments after protein digestion by trypsin. PMMA generated enough EOF for separation of peptide fragments. Encapsulated trypsin showed increased stability during continuous use than that in free solution.

Another important application of biodoped ceramics is in the field of chromatography. Monolithic silica columns are well studied for chromatography applications (Tanakaa *et al.*, 2002). Protein-encapsulating silica sol-gel columns have been successfully used in capillary electrochromatography (CEC) (Kato *et al.*, 2004; Sakai-Kato, 2003). Kato *et al.*, 2002, encapsulated two chiral compounds, BSA and ovomucoid (OVM) from chicken egg white, within a TMO-based hydrogel. Encapsulated proteins retained their activity and such prepared monolith columns showed good enantioselectivity for some enantiomers.

Biodoped ceramics are being explored for controlled delivery of drug molecules, proteins, growth factors, etc. Silica xerogel is a promising control release agent (Roveri *et al.*, 2005; Radin *et al.*, 2002). Selective intake and release of protein by silica sol-gel has been demonstrated (Rao and Dave, 1998). Biodoped ceramics can also be used as controlled release agent as shown by Santos *et al.*, 1999. Sol-gel-synthesized silica was doped with a trypsin inhibitor and it was seen that when immersed into solution, the proteins were released out in a time-dependent manner. Possible application of this protein-releasing matrix is as a substrate for bone growth. Sol-gel-synthesized silica has also been used a controlled release agent for antibody. The encapsulated antibody had retained its antibacterial properties (Radin *et al.*, 2001). Biologically active growths factor have been succesfully incorporated and released from sol-gel-derived silica (Nicoll *et al.*, 1997). Pancreatic islets-doped silica gels have been shown to be useful for both *in vitro* and *vivo* measurements for sustained, long-term regulation of insulin and blood sugar in diabetic patient (Pope, 1997).

SUMMARY

Entrapment of biologicals into ceramic matrices is a very active field. Various theoretical studies have predicted increased stability for proteins upon encapsulation in nanopores. The sol-gel technique is the most extensively used technique to achieve encapsulation of biological entities within a ceramic matrix. Several bio-friendly modifications in sol-gel routes have enabled higher stability and performance of such biodoped ceramics. Requirements for successful bioencapsulation include structural, thermal and chemical stability, functionality, and accessibility of biologicals upon encapsulation. Sol-gel synthesized ceramic matrices encapsulated biologicals in most cases retain their biological functionality and show improved stability. Diffusion difficulties for analyte molecule through a sol-gel matrix affect the reaction kinetics of entrapped biologicals. The diffusion of analytes through a sol-gel matrix depends upon the charges carried by both the sol-gel matrix and the analyte. The structural, thermal, and chemical stability depends on various factors such as the properties of the sol-gel matrix such as surface charges, porosity, ageing or drying conditions, etc and the properties of biological such as size, shape of protein or bacteria, etc. Unfortunately, there is no unifying theory available as yet which may predict the effects of encapsulation on the properties of encapsulated biologicals. Prominent applications of such biodoped ceramics include biosensors, bioreactors, cell transplant, chromatography, and controlled release agents.

Chapter Four

ELECTROSPINNING — A Novel Nanomanufacturing Technique for Hybrid Nanofibers and their Non-Woven Mats

4.1 INTRODUCTION

Conventional fiber synthesis techniques of wet, dry, melt, and gel spinning, are capable of producing polymer fibers with diameters down to the micrometer range. Electrospinning, also known as electrostatic spinning, is a process capable of producing polymer fibers with nanoscale diameters. The ultra fine solid fibers are notable for small diameters, large surface area to volume ratio, and small pore size. Fiber properties such as strength, weight, porosity, and surface functionality depend on the specific polymer used. The fundamental idea of electrospinning dates back more than 70 years ago. A series of patents describing an experimental setup for the production of polymer threads using an electrostatic force were awarded to Formhals between 1934 and 1944 (Formhals, 1934). Larrondo and Manley, 1981, first revived the technique by electrostatic spinning of polymer melts in 1981. There was little interest in electrospinning or electrospun nanofibers until mid-1990s. Since then over a hundred synthetic and natural polymers were electrospun into fibers with diameters ranging from a few nanometers to micrometers (Dzenis, 2004). It has been approximated that nanofibers with a diameter of 100 nm have a ratio of geometrical surface area to mass of $100 \, m^2/g$ (Frenot and Chronakis, 2003). The very large surface area to volume ratio, flexibility in surface functionalities, superior mechanical performance, and versatility of design are some of the characteristics that make the polymer nanofibers optimal candidates for many important applications such as composites, protective clothing, catalysis, electronics, bio-medicine including tissue engineering, implants, wound dressing membranes, and drug delivery, vessel engineering, also filtration, and agriculture. The cost efficient electrospinning technique also provides the capacity to lace togethera variety of polymers, fibers, and particles to produce nonwoven nanocomposite membranes. In the review of electrospinning by Dzenis, 2004, synthesis of composite

Nanomaterials for Chemical Sensors and Biotechnology by Pelagia-Irene (Perena) Gouma
Copyright © 2010 by Pan Stanford Publishing Pte Ltd
www.panstanford.com
978–981-4267–11-3

nanofibers was declared to be one of the biggest breakthroughs in the field reported to date. The method so widely explored with polymer solutions has been translated into forming composite nanofibers. Small insoluble particles can be added to the polymer solution or encapsulated in the dry nanofibers. Soluble drugs, bacterial agents, and metal oxide sol gel solutions can be added to the polymer and electrospun into nonwoven mats. Bringing materials into the nanometer range has been observed to not only improve material properties, but also to create new advanced characteristics not observed for bulk materials. This phenomenon is expected to advance the versatile electrospinning method onto fields where creating one-dimensional nanocomposites of ceramics, semiconductors and biologicals is of great value.

4.2 THE ELECTROSPINNING PROCESS

4.2.1 Equipment

A typical electrospinning set up consists of a high voltage power supply (Gamma High Voltage), a programmable syringe pump (Kd Scientific), syringe, needle (Popper & Sons), and a grounded collector screen. During the process a polymer solution or composite mixture is injected from a small nozzle under the influence of an electric field as high as 30 kV. The build up of electrostatic charges on the surface of a liquid droplet induces the formation of a jet. The jet is subsequently stretched to form a continuous fiber. Before it reaches the collecting screen the solvent evaporates or solidifies. The fibers are collected on a conductor surface, and form nonwoven mats.

A conventional electrospinning set up is presented in Fig. 4.1. The advantage of the vertical system is allowing the polymer fluid to drop with help of gravity

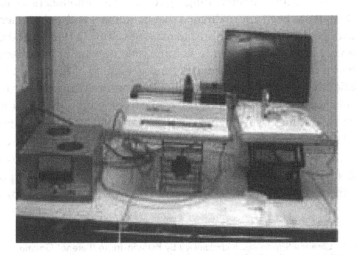

Figure 4.1. Typical electrospinning set up.

onto the collector underneath (Rekener's group, 2001). Some groups report the capillary to be tilted at a defined angle to control the flow (Buer *et al.*, 2001; Mac-Diarmid *et al.*, 2001), but introduction of the programmable syringe pump ensures a very specific controlled flow rate can be achieved. To date there has been no systematic study comparing the fibers resulting from vertical to horizontal set ups. Neither there is a report thoroughly comparing the effects of changing polarity on the resulting mats. The high voltage electrode can be inserted either in the polymer fluid or attached onto the tip of a metal needle.

4.2.2 The Process

Composite nanofibers can be obtained two experimental techniques. The first procedure known as polymer template method, utilizes electrospun polymer fiber mats to be immersed into the composite solution. The soaking process allows composite particles to be adsorbed onto the polymer fibers. It is followed by a thermal or chemical treatment to obtain the desired product. There is no control over the amount of material adsorbed into the fibers. Long processing takes up to 36 hours for completion, and is another fall back of this method.

The more versatile technique of forming composite nanofibers is that of electrospinning polymer and composite solution mixtures. This second method has been more widely reported as a tool to prepare various composite fibers. Electrospinning combination of polymers with composites allows control over composition, reduces the preparation time, and permits the use of hydrophilic polymers. The desired material is obtained through thermal or chemical treatment of composite nanofiber mats.

Rangpukan and Reneker, 2001, have reported the development of electrospinning from molten polymers in vacuum. Electrospinning from polymer melts in vacuum environment is advantageous because higher electric field strength over large distances and higher temperature can be used than in air. Multi jet electrospinning devices have been proposed by Gupta and Wilkes, 2003, and more recently by Ding's group (2004). The latter designed a movable multi-jet device that would electrospin onto a rotating grounded tubular layer to ensure uniform thickness of the electrospun membrane. Many set up adjustments, presented in Table 4.1, aim at improving fiber alignment. It is expected that simple and reproducible alignment of nanofibers can revolutionize existing applications and help develop entirely new ones.

A novel approach to position and align individual nanofibers on a tapered and grounded wheel-like bobbin has been described by Theron *et al.*, 2001. The tip edge substantially concentrates the electrical field so that as spun nanofibers are almost all attracted to and can be continuously deposited on the bobbin edge of the rotating wheel. It has been demonstrated that with this approach polyethylene oxide nanofibers with diameters ranging from 100 to 400 nm were in alignment with a distance between two fibers varying from 1 to 2 μm. The separation distance variation is due to varying repulsive forces related to nanofiber diameters and residual

Table 4.1 Set-up adjustments.

Innovation	Reference	Effect
Vacuum System	Rangpukan, & Reneker, 2001	Increased electric field strength
Multi-jet System	Gupta & Wilkes, 2003	Electrospinning various materials simultaneously, and uniform thickness
Wheel-like Bobbin Collector	Theron *et al.*, 2001	Aligned nanofibers with 1–2μm gaps between individual threads
Two Separated Electrically Conductive Substrates	Xia's group, 2003	Uniaxially aligned nanofibers stretched across the substrates
Metal Frame Collector	Dersch *et al.*, 2003	Oriented nanofibers
Insulated Mandrel Collector	Sundaray *et al.*, 2004	Aligned nanofibers
Copper Wire Framed Drum Collector	Katta *et al.*, 2004	Nanofibers oriented perpendicularly to the copper wires
Spinneret Containing Two Coaxial Capillaries	Li and Xia, 2004	Hollow nanofibers, nanocomposites of "unspinable" materials

charges. The linear velocity at the tip of the wheel collector was 22 m/s when the aligned fibers were collected.

Preparation of uniaxially aligned arrays was achieved with use of a collector consisting of two pieces of electrically conductive substrates separated by a gap whose width could be varied from hundreds of micrometers to several centimeters. The charged nanofibers driven by electrostatic interactions were observed to stretch and span across the gap at the same time became uniaxially aligned.

Dersch *et al.*, 2003 demonstrated the orientation of polyamide nanofibers on a metal frame as the collector. The observed orientation is believed to be due to the polymer jet jumping back and forth from one side of the frame to the other, apparently because of electrostatic charging effects. Sundaray *et al.*, have shown that as spun fibers could be aligned when an insulated cylinder attached to the axle of a DC motor is used as the substrate rotated with a high speed of approximately 2000 rpm.

Katta *et al.*, 2004 used a copper wire framed drum as the collector. During the process many of the nanofibers deposited directly on the copper wire closest to the needle. As the drum rotated, the next copper wire was attracting the nanofibers. It was observed that the nanofibers stretched perpendicularly to the copper wires to span across the gap between them. The rotation rate of the drum was held constant at 1rpm.

The electrospinning set up has also undergone adjustments to allow easy nanocomposite fiber formation. A spinneret containing two coaxial capillaries was fabricated by inserting a polymer coated silica capillary into a stainless steel needle (Li and Xia, 2004). Two viscous liquids were simultaneously fed through the inner core and outer sheath tubes. The core transported a heavy mineral oil, and an ethanol solution of polyvinylpyrrolidone (PVP) and titanium tetraisopropoxide flowed through the sheath. The oil was extracted from the composite fiber mats by immersing in octane overnight. Hollow fibers made of pure titania were obtained by calcination of the as spun fibers at 500°C for one hour. This experiment further developed a method of coaxial electrospinning of continuously coated and hollow nanofibers (Sun, 2003). It proved the use of nanofluidic channels a successful method to allow liquids with low electrical conductivities, like mineral oil, to be stretched into thin filaments by co-electrospinning with an easy to spin polymer solution.

4.2.3 Jet Modeling

The interest in behavior of liquid jets in presence of an electric field dates back to the work of Rayleigh (Strutt and Dublin, 1882). Taylor (1965;1969) reported experimental evidence of the conical shape projection from which a jet leaves the surface of a liquid drop. He showed that fine jets of various liquids could be attained when a semi-vertical cone angle attains 49.3°. Taylor derived an expression of the critical voltage Vc in terms of the distance between the electrodes (H), length of the capillary tube (L), radius of the tube (R), and surface tension of the fluid (γ)

$$Vc^2 = 4(H^2/L^2)(\ln(2L/R) - 1.5)(.117 * \pi * R * \gamma) \tag{4.1}$$

Hendricks defined the minimal spraying potential of a conducting drop suspended in air, where r is the jet radius (Hendricks *et al.*, 1964).

$$V = 300(20 * \pi * r * \gamma)^{1/2} \tag{4.2}$$

The required voltage will be lower if the process is performed in vacuum due to the increased electric field (O'Konski and Thatcher, 1953).

The electrospinning community has focused on the properties of liquid jets emitted from the Taylor cone. Polymer solutions are highly viscous and exhibit nonlinear rheologic behavior. The surface tension appears to be a function of solvent composition, and negligibly dependent on the polymer (Fong and Reneker, 1999).

A model of weakly conductive viscous jet motion accelerated by an external electric field was formulated taking into account inertial, hydrostatic, viscous, electric, and surface tension forces. The model demonstrated that the Taylor cone does not represent a unique critical shape of 49.3°. It proved half angles to be much closer to the shape of the jet. Elastic force affects morphologies of the electrospun fibers for liquids with a non-relaxing elastic force (Yarin and Reneker, 2001).

It was observed by Reneker's group (2000;2001) that the bending instability develops due to the mutually repulsive forces resulting from the electric charges of the jets. Capillary instability, resulting from surface tension, can be prevented by the strong stabilizing influence of viscoelastic stresses. There is a tendency of the charges in the polymer fluid to reduce their Coulomb interaction energy by moving the fluid in a complicated path.

To calculate the bending electric force on a polymer jet a localized approximation was developed. Partial differential equations were derived and used to predict the growth rate of small electrically driven bending perturbations of a liquid column, accounting for solvent evaporation and polymer solidification. The method was also used to calculate the jet paths, during the course of nonlinear bending instability leading to formation of large loops and resulting in nanofibers (Yarin *et al.*, 2001).

Hohman, Shin, Rutledge and Brenner (2001) also studied electrospinning, with regard to electrically forced jets and instabilities, and proposed a stability theory. The authors analyzed mechanics of the unstable — "whipping" jet by studying the instability of an electrically forced fluid jet with increasing field strengths. The electrical instabilities are enhanced whereas the Rayleigh instability is suppressed at increasing field strengths. Dominating instability strongly depends on the surface charge density and the radius of the jet. The stability theory was then used to develop a quantitative method for predicting when electrospinning occurs.

A later model suggests that there is a limiting diameter for the fluid jet attained during electrospinning. The model predicts that the final diameter of the fluid jet arises from a force balance between surface tension and electrostatic charge repulsion. The 2/3 scaling with the inverse volume charge density predictions quantitatively agrees with electrospun fibers produced from solutions of different polymers at various concentrations (Friedrich *et al.*, 2003).

Using Monte Carlo simulation technique static and dynamic properties of amorphous polyethylene nanofibers were studied (Vao-Soongnern, 2000). These workers concluded that fibers of different sizes exhibit almost identical hyperbolic density profiles at the surface, and that the mobility of the polymer chain in a nanofiber increases as the diameter of the nanofiber decreases.

4.2.4 *Solution and Process Parameters*

Theron *et al.*, 2004, reported a systematic investigation of the effect variation of the governing parameters has on the electrospinning of polyethylene oxide (PEO), polyacrylic acid (PAA), polyvinyl alcohol (PVA), polyurethane (PU) and poly-caprolactone (PCL) solutions. It was found the volume charge density decreases with increasing flow rate for all the solutions tested. The surface charge density decreased with increasing flow rate and voltage applied. The electric current in the jet increased with flow rate for PEO, PVA, PAA ($2.5*10^5$) and PU, but decreased for PCL. The volumetric flow rate in the higher molecular weight solution of PAA had no effect on the electric current. At increased values of the volumetric flow rate the faster solution discharge at the tip of the needle limits the solution's exposure time to the electric charge applied. The fact that the volume charge density decreased exponentially as the nozzle to ground distance increased was explained by the decrease in electric field strength. Addition of ethanol to an aqueous PEO solution decreased the volume charge density when other parameters were kept constant, achieving higher evaporation rate, which facilitates nanofiber solidification.

Deitzel *et al.*, 2001, investigated the formation of beads and found out that the spinning voltage influences mainly the formation of beads while the polymer concentration has effect on the fiber size. Fiber diameter increased with increasing polymer concentration according to a power law relationship. At high concentrations a bimodal distribution of the fiber sizes was observed.

Wilkes' group (2003) has studied the effect of solution concentration, needle-screen distance, electric potential at the tip, and flow rate on the electrospun nanofibers. Bead structures appeared when the needle to collector distance was decreased with an increase in the average fiber diameter. Higher concentration solutions formed fibers of increased average diameter. Bead like structure was observed to turn into blobs at lower capillary-screen distance. Finally, increasing the potential decreased the fiber diameter. A lower and upper limit of polymer concentration was determined. High concentration led to failure as the viscosity was too high. Low concentration lacked enough viscosity, and resulted in a high flow rate.

The electrospinning solution must have a polymer concentration high enough to cause entanglements with viscosity low enough to allow motion induced by the electric field. To prevent the jet from collapsing into droplets before solvent evaporates the surface tension must be low. Morphological changes can occur upon decreasing the distance between needle and the substrate. The increase of needle to collector distance or a decrease in the electrical field will result in reduced bead density, regardless of the polymer solution concentration. Applied electric field can influence the morphology in periodic ways, creating a variety of new shapes on the surface (Fong *et al.*, 1999).

Demir *et al.* (2002) found that the fiber diameter was proportional to the cube of the polymer concentration. While electrospinning polyurethane nanofibers, the authors recognized that the fiber diameters obtained from the polymer solution at a 70°C temperature were much more uniform than those obtained at room temperature. The viscosity for same concentration solutions at higher temperature was noted to be significantly lower than that at room temperature. Increasing the electrospinning solution temperature controls the morphological imperfections such as beads or curly fibers. The increase of electrical potential resulted in rougher nanofibers.

Zong *et al.* (2002) observed that 1wt% salt introduced to biodegradable poly(D,L-lactic acid) polymer solution resulted in bead-free nanofibers. Addition of salt raised the charge density brining more electric charges to the jet.

A comparison of electrospun surface using AC potential to corresponding structures using DC potentials was done (Kessick *et al.*, 2004). It was demonstrated

Table 4.2 Electrospinning Process Parameters.

Needle to Collector Distance	Exponentially inverse to the volume charge density Inversely proportional to bead formation density Inverse to the electric field strength Inversely proportional to fiber diameter
Flow Rate	Directly proportional to the electric current Directly proportional to the fiber diameter Inversely related to surface charge density Inversely related to volume charge density
Voltage	Inversely proportional to surface charge density Direct effect on bead formation AC potential improved fiber uniformity Inversely related to fiber diameter

Table 4.3 Electrospinning Solution Parameters.

Concentration of Polymer	Directly proportional to the fiber diameter Power law relation to the fiber diameter Cube of polymer concentration proportional to diameter Parabolic — upper and lower limit relation to diameter
Ionic Strength	Directly proportional to charge density Inversely proportional to bead density
Solvent	Effects volume charge density Directly related to the evaporation and solidification rate
Temperature	Inversely proportional to viscosity Uniform fibers with less beading
Viscosity	Parabolic relation to diameter, and spinning ability

that AC potential significantly increased the surface coverage on insulating substrates and reduced the amount of fiber whipping inherent in the DC process.

Nanosized beads and nonwoven porous fiber constructs of poly(E-caprolactone) were produced by electrospinning (Hsu and Shivkumar, 2004). Nearly spherical beads with diameters between 900 nm and 5 μm were produced with dilute solutions with less than 3wt% PCL. An upper and lower limit for solution concentration during electrospinning was observed, it is said to correspond to the level of entanglement between the polymeric chains. The beads formed when the jet at the end of the Taylor cone split into many mini-jets and each mini-jet disintegrated into small droplets. The fibers were formed when the jet at the end of the Taylor cone undergoes continuing, extensional flow and then disintegrates into several stable mini-jets that accelerate toward the collector.

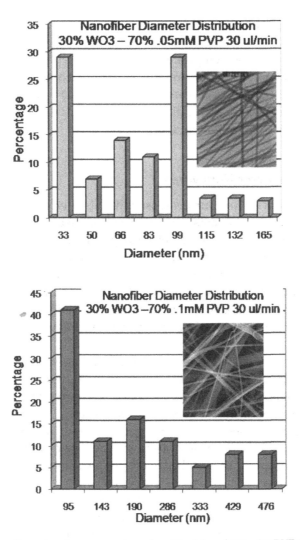

Figure 4.2. Fiber diameter size comparison for .05 mM and 0.1 mM PVP-WO$_3$ composite nanofibers electrospun at equal metal oxide ratios, and constant volumetric flow rate.

There has been only one systematic study of composite fiber diameter variations. The nanocomposite fiber thickness was reported to be directly proportional to the process flow rate, and polymer concentration. Figure 4.2 compares the nanofiber diameter distribution for 0.05 and 0.1 mM polyvinylpyrrolidone (PVP) solution mixtures of equal ratios to tungsten oxide sol-gel, and constant flow rate. The diameter was reported to be inversely proportional to the amount of metal oxide used. Finer diameter fibers obtained at lower flow rates, and lowered solution viscosity reported in (Sawicka *et al.*, 2005) can be justified by extending the rules discussed for pure polymer electrospun fibers. The consistent result observed by all groups who have electrospun nanocomposite fibers is that the thread diameter decreases after chemical or thermal polymer extraction.

4.3 APPLICATIONS

Most of the polymer, and composite nanofiber applications have not yet reached industry application level, but are still at a research and development stage. The four main categories of research reported for pure polymer and composite nanofibers are bio-medical applications, semiconductor nanowire synthesis, structural composite materials, and filtration.

4.3.1 Semiconductor Nanowires

The next category of electrospun composites has also tremendously expanded possible applications for nanofibers. One-dimensional metal oxides have been extensively researched due to improved electro-optical, electro-chromic, ferroelectric, catalytic, and gas sensing properties. The large aspect ratio of surface area to volume observed for electrospun mats imply improvement of adsorption, and reaction rates. Versatility of design, and the straightforwardness of electrospinning promise its implementation in multi-functional applications. The first experiments conducted in this field involved the multi-step fiber template technique described previously.

Bognitzki *et al.* (2000) reported coating of electrospun poly(lactic acid) (PLA) fibers to manufacture polymer-metal hybrid nano and mesotubes. Selective removal of organic component from blended fibers resulted in fine porous and functional nanofibers with specific surface topologies.

The above experiment was shortly followed by Caruso *et al.* (2001) who coated poly(L-lactide) electrospun fibers with amorphous titanium dioxide sol-gel. In this report hollow titania fibers were produced by removal of the thermally degradable polymer. The sol-gel coating was observed to mimic the finer details of the nanofibers like nodules formed on the inner walls of the tubes.

The next experiment used electrospun fiber membranes as substrates for liquid-phase deposition (Drew *et al.*, 2003). The continuous coatings of titanium dioxide (TiO_2) took 12–36 h, and tin dioxide (SnO_2) took 12–28 h to prepare. The metal oxide covered the fibers completely, and coated without binding them together. Additional spherical tin dioxide particles formed on the fiber surface and next to the fiber, the authors believed it to be a result of the nucleation process.

Most recent publication utilizing similar procedures reported formation of vanadium oxide nanofibers through polylatide (PLA) fiber coating with metal sol-gel followed by a hydrothermal treatment (Whittingham *et al.*, 2005).

The technique of direct blend electrospinning came along few years later, possibly due to the more complicated formula of electrospinning polymer with composite mixtures. It became far more popular in the synthesis of metal oxide nanofiber synthesis than the fiber template method.

The first report of electrospinning metal oxide composite fibers was made in 2002 by Dai *et al.*, Alumina borate solution was mixed with polyvinyl alcohol (PVA) to form a viscous gel that was electrospun. Composite nanofibers were calcinated above 1000°C to form pure $Al_4B_2O_9$, $Al_{18}B_4O_{33}$, and stable phase of α-Al_2O_3 was reported to form at 1400°C.

Titanium dioxide was the next metal oxide electrospun into composite nanofibers (Li and Xia, 2003). Titania sol-gel was directly added to an alcohol solution containing polyvinylpyrrolidone (PVP) and electrospun to form composite nanofibers. Pure metal oxide nanofibers were achieved by a heat treatment at 500°C in air for 3 h. Authors reported an upper and lower limit for the electric field applied and its effect on the jet stability. They observed the fiber diameter to be proportional to the polymer concentration, and feeding flow rate. Attachment of gold or other metal nanoparticles to TiO_2 cause the Fermi level of titania to be shifted to more negative potentials and prevent the recombination of electron-hole pairs, improving photocatalytic activity and photoelectrochemical response. The group employed the photocatalytic reduction of $HAuCl_4$ in the presence of organic capping reagents to selectively deposit gold nanostructures on electrospun anatase nanofibers (Li *et al.*, 2004). Another group combined the methodology for forming mesoporous TiO_2 with the technique of electrospinning to produce mesoporous titanium dioxide fibers (Madhugiri *et al.*, 2004).

Most recently long pure titania nanofibers and those modified with erbium oxide were fabricated by electrospinning followed by thermal pyrolysis (Tomr *et al.*, 2005). The reported material can function as a selective emitter material for thermophotovoltaic applications. The procedure described for alumina borate, and titania fibers has been extensively used to synthesize various metal oxide nanofibers.

Dharmaraj and Viswanathamurthi (2003; 2004) have reported numerous metal oxide nanofibers prepared by electrospinning. They have successfully electrospun polyvinyl acetate (PVAc) with vanadium sol-gel to create composite nanofibers, and calcinations of as received membranes resulted in pure vanadium pentoxide nanofibers. Similar procedure was followed for magnesium titanate, p-type semiconductor palladium oxide, nickel titanate, and ruthenium doped titanium dioxide.

Shao's group (2003–2004) has electrospun polyvinyl alcohol (PVA) mixed with various metal oxide solutions to create metal oxide composite nanofibers, followed by calcinations of precursor membranes to result in pure metal oxide nanowires. To date their group has successfully synthesized Co_3O_4, NiO, CuO, Mn_2O_3, Mn_3O_4, ZnO, ZrO2, NiO/ZnO, $NiCo_2O_4$, CeO_2, and $LiMn_2O_4$ nanofibers.

The common observation for all cited articles of metal oxide nanofiber formation is the polycrystalline grains along the decomposed polymer frame morphology of the heat-treated composite fibers. In recent years, the author's group reported single crystal tungsten oxide nanowires (Gouma *et al.*, 2005–2008). The synthesis procedure of novel polymer-oxide composite nanofibers by means of electrospinning, focusing on the PVP- MoO_3 / WO_3 systems followed the steps of the blend mixture electrospinning. Postheat treatment samples contained pure metal oxide nanowires without organic material present. The TEM study manifested the same diffraction pattern along the metal oxide nanowire

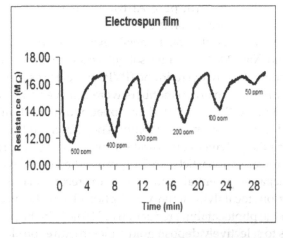

Figure 4.3. Sensing data for the nanofiber-based WO$_3$ film's response to NO$_2$; TEM micrograph of fibers consisting solely of crystalline WO$_3$ (see inset diffraction pattern).

as presented in Fig. 4.3. Our previous work of sol-gel processed thin films of WO$_3$ tested their sensitivity for NO$_2$ and NH$_3$. Tungsten oxide thin film sensors were found to be highly specific to NO$_2$. Therefore, the nanofiber structures were employed as NO$_2$ sensing material, and compared to the response observed for thin films of the same material.

The electrospun composites showed significantly higher gas sensitivity than the thin films. The detection limit was shifted to lower gas concentrations below 50 ppm, and the response time was reduced from minutes to seconds. Another benefit observed in using nanowires was the improved stability of the sensor, correction of baseline drift.

Metal oxides represent majority of the work reported on electrospinning nanocomposites. Another group of semiconductors as composite nanofibers promises to become the next milestone in the electrospinning technique is silica. The first attempt to electrospin silica fibers dates back to 2003 (Balkus *et al.*). Non-woven mesh of mesoporous molecular sieve fibers in the micrometer range was reported. These workers observed substrate conditions that promote condensation of silica improved fiber formation. Very recently SiO_2 nanofibers have been created by two different approaches: direct spinning of silica precursors and spinning of polymer nanofibers followed by sol-gel silica coating of nylon-6 nanofibers (Zhang *et al.*, 2005). The directly spun SiO_2 nanofibers are typically smoother than those produced by the sol-gel coating method.

4.3.2 Bio-Sensing

Our group has utilized the electrospinning process to create enzymatic biocomposite nanofibers (Sawicka *et al.*, 2004). This experiment for the first time connected the fields of electrospinning and biosensing by employing urease nanofiber mats as novel urea biosensing material (see Fig. 4.4). The mechanism used in enzyme-based urea biosensors concentrates on irreversible hydrolysis of urea into ammonia and carbon dioxide in the presence of urease. Blends of polyvinylpyrrolidone (PVP) and urease were electrospun, and formed beaded nanofibers with diameters of 7 to 100 nm. The encapsulated enzyme activity was comparable to its pure form. Further studies to be reported elsewhere found solvent substitution and increased polymer concentration allowed a larger amount of enzyme to be electrospun to form smoother fibers without bead structures. The successful incorporation of urease into biocompatible polymer nanofibers expands the material applications into a portable personal blood urea analysis biosensor, an implantable patch for urea removal for kidney failure patients through peritoneal, and haemodialysis.

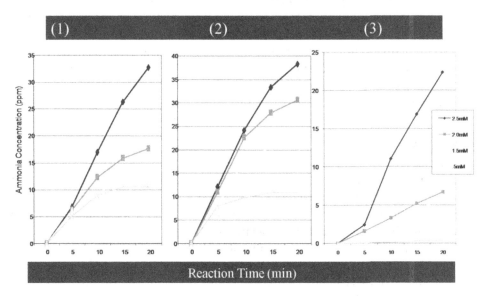

Figure 4.4. Ammonia concentration vs. time when urea solutions reacted with (1) 0.2 ml of urease in PBS buffer, (2) 0.2 ml 30% urease in buffer/70% PVP in ethanol solution, and (3) 0.1 ml of urease/PVP nanofiber mat. For color reference, turn to page 151.

The immobilized urease could also be utilized in analysis of creatinine, and arginine concentrations. Another important application of immobilized urease would be to determine heavy metal and other pollutant contaminations due to loss of enzyme's activity when introduced to trace amounts of contaminants. The successful encapsulation of urease into electrospun nanofibers promises to translate into a successful incorporation of other enzyme for a variety of applications.

4.3.3 Encapsulation of Biological Reagents

The successful incorporation of biological reagents into electrospun nanofibers, and sustained activity through the harsh environment of the electrospinning process has recently been reported. Long rod-shaped M13 viruses were electrospun into micro- and nanofibers (Lee and Belcher, 2004). The viruses were blended with polyvinylpyrrolidone prior to electrospinning to improve processing, continuous M13 virus-blended PVP fibers were fabricated at a 1:4 rationed suspension. The resulting virus composite nanofibers were transformed into nonwoven fabrics that after dissolving in 3mL of TBS the suspension were rested for infection ability of the M13 virus. The dissolved virus suspension retained ability to infect the bacterial host.

The electrospinning method was utilized to form beaded nanofibers containing photochromic pigment called purple membrane (PM) for possible applications in optical processing and optical data storage (Fisher and Hampp, 2004). A complete encapsulation of PM patches, ranging in size from 300 to 800 nm, was confirmed by absence of free bacteriorhodopsin (BR) at any of the beads. Embedding of particles significantly larger than the fiber diameter is seldom due to expected breakdown of the spinning process as microscaled particles in the polymer solution pass the nozzle. The unusual result is presumably due to the flexibility of PM patches. Electrospinning of micrometer size reagents into nanofibers may be an alternative to microencaplusation for biomolecules. The results obtained with PM patches may be applied to other biological functional macromolecules and even cell organelles.

SUMMARY

Electrospinning method is a vital tool for nanotechnology. There is no question as to the impact of this nanomanufacturing technique. The only debate is rather which of the fields currently researched will pioneer in first leap from research into production. The semiconductor composite nanofiber — nanowire formation is still in development. Process and solution parameters, thermal stability, and mechanical functionality are some of the systematic studies of composite fibers expected. Advancements of this category of materials will expand their applications respectively. Use of nanowires presents capabilities of revolutionizing electronics. It has proved to immensely improve the sensing capabilities of previously used materials. The union of nanotechnology and enzyme-based biosensing is only in its infancy, however the successful trials with urease promise to transcend into use with other enzymes, and further expand the possible purpose of electrospinning. Use of biocompatible and degradable polymers open the door into implantable monitoring devices or therapeutic patches. The latest triumphant attempt to electrospin semiconductor material formed nanofibers of silica. This promises to be the next breakthrough in the field, since SiO_2 has been extensively researched in the form of gels for optical biosensing, and other purposes. One can expect silica thin film applications to transform into electrospun silica membranes. Nanofiber membranes have also found applications in tissue engineering, as it will be discussed extensively in the next chapter, and in filtration and strengthening composite formation. All of these fields offer great prospects, and a promise for a better human health and welfare.

Chapter Five

Nanomedicine Applications of Nanomaterials

5.1 ELECTRONIC NOSES AND TONGUES

An increasing demand for electronic instruments that can mimic human olfactory processes and may provide low-cost, rapid sensory information to hasten the process of odor evaluation for applications such as food quality assessment, environmental monitoring, and even forensics led to the conceptualization of an electronic nose. This is defined as an intelligent chemical sensor array system for odor classification (Gardner and Bartlett, 1999; Kress-Rogers, 1997). Electronic nose systems, including the materials used to build the sensing elements, device architecture, and "intelligent" signal processing routines. Recently, medical applications of electronic noses have been explored. The use of a novel electronic nose to diagnose the presence of pulmonary infection and to distinguish between serum and cerebrospinal fluid, as might be encountered in exudates collected from the eye or ear, was reported (Thaler *et al.*, 2000). Osmetech has obtained U.S. Food and Drug Administration approval to use its multiple detector-based device for detection of urinary tract infections in patients and is currently seeking approval for use of the device in diagnosis of bacterial vaginosis (http://www.osmetech.co.uk/enose.htm).

Clearly, this field is still in its infancy, and the opportunities for developing improved biosensors are abundant. The diverse nature of their applications appeals to the public in general. It is envisioned that handheld electronic nose/tongue devices will be the health monitoring technology of the future, whereas through a simple exhaled breath or nasal expired air analysis physiological functions may be monitored at all times. At the same time, novel materials such as chemoselective polymers and bio-nanocomposites, carbon nanotubes, and oxide nanobelts and nanowires have revolutionized the field of bio-/chemical sensing(Novak *et al.*, 2003; Comini *et al.*, 2004; Wan *et al.*, 2004; Gouma's group, 2006). The high surface to volume ratio of nanostructured materials favors gas adsorption on these surfaces thus enhancing the sensitivity of the sensor. Modifications in the electronic structure of nanoscale semiconductors can affect the optimum temperature for gas sensing. By employing these new materials, advanced

Nanomaterials for Chemical Sensors and Biotechnology by Pelagia-Irene (Perena) Gouma
Copyright © 2010 by Pan Stanford Publishing Pte Ltd
www.panstanford.com
978–981–4267–11-3

sensing systems are being developed that are faster, more selective, and highly sensitive to harmful chemical species and to "disease-signaling" gases. The issue of chemical selectivity of each sensor component is currently being compensated for through the use of pattern recognition algorithms and neural network routines that process the signal from nonselective sensing elements and define a spatial distribution of the electronic nose responses to different types of chemicals.

Traditionally, gas analysis has relied on gas chromatography and mass spectroscopy (GC/MS) systems. Gas Chromatography and mass spectroscopy are used to identify and quantify volatile and semivolatile organic compounds in complex mixtures. Components of a chemical mixture are separated in the GC and identified by their respective masses in the MS. Organic compounds must be dissolved in volatile and organic solvents for injection into the GC. The duration of a gas chromatographic run is between 20–100 min, and this is the instrument analysis time. GC-MS may identify unknown organic compounds by matching spectra collected with reference spectra, or by a priori spectral interpretation. Data analysis can take over 20 hr. Such analytical systems, although accurate in identifying volatile organic compounds, are expensive and bulky, and require experienced operators; whereas electronic noses and tongues offer the promise of fast, reliable, portable gas and/or liquid detection systems for user-friendly operation. Microfabricated sensor arrays are currently built in small, hand-held devices. Each sensor response varies from seconds to minutes. Data processing may be simplified to provide a straightforward reading of the gas chemistry and concentration, or the quality of the odorous mixture.

This section examines the current status of E-nose technology to obtain insight on how this field will evolve due to nanotechnology.

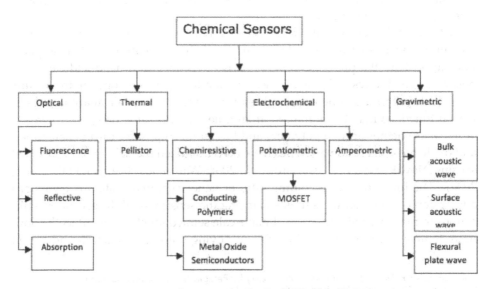

Figure 5.1. Sensor technologies utilized in E-noses.

5.1.1 State-of-the-art

Electronic noses (gas phase odor detection instruments) and tongues (their liquid phase detecting equivalents) rely on several different chemical sensor materials, such as metal oxides, semiconductors, polymers, dyes, etc.

5.1.1.1 Metal Oxide Sensors

Semiconducting oxides have been the preferred low-cost sensing elements for the detection and monitoring of permanent gases (such as CO). They show high gas sensitivity, fast response to the presence of the gaseous analyte, and good long-term stability. The detection process of oxidizing/reducing gases by semiconducting metal oxides involves the change in oxide conductivity in the presence of the gas due to catalytic reduction/oxidation reactions occurring at the oxide surfaces (Moseley and Crocker, 1996). These catalytic reactions are controlled by the electronic structure of the oxide system used, as well as by the chemical composition, crystal structure, and relative orientation of the surfaces of the oxide phase(s) exposed to the gas (Gouma, 2003).

There has been evidence in the literature of selective detection of a particular gaseous analyte in the presence of interfering gas mixtures (i.e., sensor selectivity) (Gouma's group, 2003–2008) that is largely determined by the chosen crystalline polymorph (specific crystallographic phase) of a stoichiometric and pure metal oxide used for sensing. For example, CO and hydrocarbons are sensed by rutile-type structure (Dutta *et al.*, 1999) such as the polymorphs of SnO_2 and TiO_2 while oxides with perovskite structure may be used to sense oxidizing gases with higher sensitivity. Because a given crystal structure may be sensitive to more than one gas, sensing tests at different temperatures are typically carried out so as to identify the optimum operating temperature for the specific sensor (Semancik and Cavicci, 1993). It is important to remain within the phase stability field of the particular polymorph of the oxide to attain reliable and reproducible sensing properties.

There is an increasing trend in chemical sensing to utilize nanostructured oxides, such as SnO_2 nanobelts, as gas sensing elements (Comini *et al.*, 2004). Nanocrystalline processing may also be used to stabilize oxide polymorphs that would otherwise be energetically unfavorable under normal testing conditions, such as anatase-phase of TiO_2, as opposed to its stable rutile phase (Gouma *et al.*, 1999). For resistive-type chemical sensors, it was observed that the sensing properties, such as response time and gas sensitivity, appeared to improve when the size of the oxide particles was reduced. The observed specificities of the different metal oxides for different gases, combined with the rapid detection of very small quantities of the analyte gases suggest that these materials may be used as the components of the next generation of chemical and bio-sensing electronic noses. Furthermore, the doping of oxides with biological molecules (enzymes, cells, antibodies) — as discussed in Chapter 3) provides a feasible strategy for forming sensing elements specific to certain pathogens. This alternative approach for design of biosensors, i.e. the intrinsic specificities of the enzymes would provide the

Figure 5.2. SACMI's E-nose is based on 6 SnO_2 sensors. For color reference, turn to page 151.

means of identification of the substrates, and the need for an array of multiple detectors with limited specificities along with elaborate pattern-recognition electronics would thereby be obviated.

5.1.1.2 Polymer-Based Sensors

The most common gas sensing elements rely on sorption-based detector materials, such as conducting polymers or conductive polymer composites (Sisk and Lewis, 2003). The swelling of polymers due to adsorbed chemical species can change the electrical properties of conductive polymers, as well as the oscillation frequency of polymer-coated cantilever devices. Both conductivity and oscillation frequency can be used as analytical parameters; conductivity is discussed here, while oscillation is covered in the "Other Sensor Materials and Technologies" section below. The principle of gas detection by conductivity changes is the adsorption of the volatile analytes on composites consisting of a conductive matrix blended with polymers for which the analytes have variable affinities (Mather, 2002; Stetter *et al.*, 2000; Fisher-Wilson, 2001). As the analytes adsorb to the polymers, the dimensions of the composites change slightly, resulting in small but detectable changes in conductance. Various polymers, especially polyheterocycles such as polypyrroles, have been employed for their capacity to bind volatile analytes. Although there are clearly differences in the affinities of a number of gas phase analytes for these polymers, there is little absolute selectivity in the absorptive process on which the detectors depend for recording conductance changes. The responses of these sensing materials depend on their molecular volume, branching of the polymer chain, hydrogen bonding, etc. It is not clear yet what is the effect of each of these parameters to gas selectivity. The absorptive process has been likened to the association of volatiles with an organic solid phase in gas-liquid chromatography. Even with the best-designed composites, the kinetics of adsorption and desorption of the volatile analytes have half-lives on the order of hundreds to thousands of seconds which suggest relatively long gas detection times.

Because the composites respond to adsorbed volatile analytes by changes in dimensionality that lead to altered conductance, the choice of polymers may be compromised by the need to achieve a conducting composite. One strategy employs polymers with some intrinsic conductance (several of the polyhetero-cycles have this property and are therefore favored in design of composites): this approach limits the range of adsorptive selectivity. Another approach (She-vade *et al.*, 2003) employs nonconductive polymers [such as poly(4-vinylphenol), polyethylene oxide, ethyl cellulose] blended with carbon black, which serves as the conducting component of the composite. Both strategies result in detectors that are most sensitive to volatiles with a relatively high vapor pressure, such as fatty acids and related alcohols, rather than permanent gases such as methane, NO_2, or CO. Some of the relatively high vapor pressure organic acids and alcohols are products of bacterial or yeast fermentation and are therefore encountered in environments such as the headspace of vessels used in wine making.

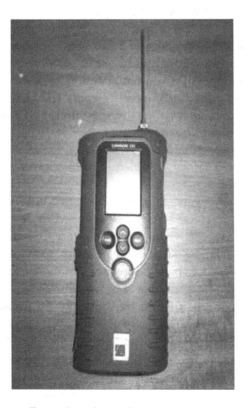

Figure 5.3. The Cyranose E-nose based on polymer sensor arrays. For color reference, turn to page 152.

A commercial electronic-nose system based on polymer-carbon black composite sensors is the Cyranose 320. This technology relies on resistive type detection mode due to the swelling of the polymer films. The detectors used in this electronic nose are nonselective. Thus, the sensor response is not correlated with the concentration of a single gaseous species (chemical compound), but is a combination of all the chemical information contained in the sample, which, in this case, is the "smell print" formed from the headspace of bacteria solutions. In order to perform identification, the current instruments must be "taught" to associate each volatile with a distinctive pattern of responses from the detectors, but when multiple volatiles are present at various levels, it may be difficult to achieve identification, even with sophisticated pattern-recognition logic.

5.1.2 Other Sensor Materials and Technologies

LibraNose is an example of an electronic nose based on eight quartz microbalance (QBM) sensor arrays coated with metalloporphyrin compounds as the chemically sensitive materials (DiNatale *et al.*, 2003). The operation principle relies on the variation of the fundamental oscillating frequency (Df) of a thin quartz crystal as a result of the adsorption of gas analyte molecules on its surface, which changes the oscillating mass (Dm) of the system, as described by Sauerbrey law (Ballantine *et al.*, 1997).Good sensitivity was obtained with this system for aromatic compounds, amines, alcohols, and ketones. LibraNose was used for lung cancer identification by breath analysis in a study involving 60 individuals (DiNatale *et al.*, 2003). Certain volatile compounds found in the exhaled human breath of lung cancer affected individuals, mostly alkenes and benzene derivatives, have been considered as candidate markers of this disease. Di Natale *et al.* in their study had their test subjects breathe in a 4lt volume disposable bag; the sampled bags were then analyzed with the electronic nose on-site. Using multivariate analysis to process the data obtained, complete identification of the sample of diseased individuals was possible.

Polymer-coated cantilevers (e.g., microfabricated beams of silicon) have been considered as nanomechanical sensor devices used to detect physical/chemical interactions between the reactive layer on its surface (polymer film) and the environment (Lang *et al.*, 2002). Swelling of the polymer upon interaction with volatile species forces the cantilever to bend because of surface stresses when used in static mode. In dynamic mode, the cantilever acts as a microbalance driven at its resonance frequency. Changes in its mass as low as 1 pg (caused by binding reactions) change the resonance frequency of the oscillating cantilever. The addition of biochemically active layers onto the cantilever surface enables the monitoring of mass changes during molecular recognition reactions. The reliability of these sensing materials depends on the precision in microfabrication that is required to form structures with reproducibility in resonance frequencies better than a few tenths of a percent. Such cantilever array sensors (consisting of eight polymer-coated cantilevers) were used to detect acetone in exhaled air.

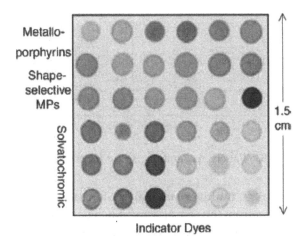

Figure 5.4. Suslick's colorimetric sensor aray. For color reference, turn to page 152.

Another electronic-nose technology involves an optoelectronic nose using colorimetric sensor arrays. These colorimetric sensors consist primarily of metallorganic compounds deposited on porous polymer membranes or other inert solid supports. Metalloporphyrins, in particular, are the sensing elements of choice in this study. The reason for their selection is that mammalian olfactory receptors are metalloproteins and most odorous compounds are excellent ligands for metal ions. Their chemical selectivity and sensitivity depend on the nature of the central metal and the peripheral constituents of the porphyrin complex (Brunick *et al.*, 1996).

Extending the use of electronic noses to liquid environments, mass sensitive devices have also been used in taste sensors or electronic tongues. These utilize primarily electrochemical detectors and voltammetric devices. Emerging trends in taste sensors include the application of spectroscopic methods in the optical tongue paradigm of Fourier transform infrared-based sensing (Edelmann and Lendl, 2002).

5.1.3 Pattern Recognition and Multivariate Chemometric Methods

Data analysis and recognition processes are not usual areas for materials scientists to focus their efforts, however these are key aspects of the electronic olfaction technology. Electronic nose data analysis correlates each tested sample to a vector in multidimensional space by means of classical nonparametric techniques. These are mathematical procedures that make no assumptions about the frequency distributions of the variables being assessed. Principal component analysis (PCA) algorithms are used to project the data sets into two dimensions (principal components). In this way, maximum distinction performance between subject classes is achieved, as well-separated clusters of measurements are projected in principal component space.

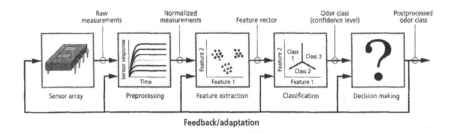

Figure 5.5. Pattern recognition process for an E-nose system.

In another approach, multivariate analysis is used to extract the maximum amount of information from the sensor array. Such parametric statistical methods assume that the distribution of the variables being assessed have certain characteristics (e.g. analysis of variance (ANOVA) assumes that the data obtained are normally distributed).

Artificial neural networks are also used for further processing of the data from the electronic noses for improving the analyte identification rate. These consist of hierarchically organized layers of information processing elements similar to the biological nervous system and have learning ability that sets them apart form the other classification methods. There may be different levels of complexity of the electronic nose and tongue subject data analysis and recognition process.

Considering the future of electronic-nose technology, there are two approaches that seem to naturally evolve. One is the use of hybrid (or orthogonal) electronic olfaction and taste systems, composed of more than one type of sensor (e.g. metal oxide-resistive and polymer composite-resistive) or using arrays of hybrid sensing elements (e.g., chemoselective membranes deposited on metal oxide sensing films). The second approach is to use small detector sets (2 to 3 sensor arrays) with high specificity, targeted for a given application. In both cases, emphasis is paid on improving the semi-selective nature of sensor materials and reducing the need for complex algorithms for signal discrimination.

Nanotechnology is expected to have a major impact on shaping the future of the fields of biochemical detection as it involves new materials, structures, and devices with improved properties. For example, novel nanomanufacturing processes, such as electrospinning (Shreuder-Gibson *et al.*, 2003), produce self-standing pure or composite material nanostructures (membranes) with high surface area for enhanced chemical attachment of analyte species and bio-catalytic processes, for on-line monitoring and/or advanced power systems (e.g. bio-fuel cells). Therefore, the electronic noses of the future are envisioned to be tiny devices that will fit in a wrist watch or a toothbrush and which will alarm us of any health problems and will protect us from exposure to allergens and pollutants. Similarly, miniaturized electronic tongues will taste the freshness of our food and purity of the water we drink. And all these will be possible at a low cost and ease of operation. And it is this improvement in the human welfare through novel technology that materials scientists are striving for.

5.2 SUMMARY

The performance of many chemical sensor systems could be greatly enhanced by accelerating the response of the sensor to the presence of analyte. Nanostructured materials provide an opportunity to significantly enhance the speed of response of a variety of sensors by virtue of their high surface area. Thus, transducers that measure mass such as resonators, would respond nearly instantaneously with nanostructured chemoselective coatings as these films would provide extremely high surface areas.

The ability to prepare high surface area chemically specific coatings, through the use of nano-manufacturing techniques such as electrospinning, is particularly important for the detection of low volatility materials such as explosives because the vapors from these materials diffuse at slow rates and therefore diffusion of the analyte into a polymer film requires too much time to meet current desired system performance characteristics. The high surface area of electrospun polymer coatings would provide a highly efficient collection surface for chemical sensing that translates into a rapid sensor response.

5.3 BREATH ANALYZERS

There have been limited reports of the use of the existing E-nose designs based on pattern recognition from multiple adsorption detectors to recognize the appearance of pulmonary infections. Use of an electronic nose to distinguish cerebrospinal fluid from serum is another application. Osmetech has obtained FDA approval for detection of urinary tract infections in patients using its multiple detector-based device and is currently seeking approval for use of the device in diagnosis of bacterial vaginosis. Clearly this field is still in its infancy and the opportunities for developing improved biosensors are abundant.

There is a need for reliable, fast and inexpensive breath gas detector systems for medical diagnostics, envisioning the development of personal monitoring devices for asthma, blood cholesterol, and even lung cancer. Although analysis of body fluids (blood, sputum, urine) for disease diagnoses and monitoring is routine clinical practice, human breath analysis methodologies that exploit the non-invasive nature of such diagnoses are still under-developed. Since the antiquity, such as the time of Hippocrates, exhaled breath was recognized as non-invasive marker for disease.

Breath testing devices first appear in 1784 when Lavoisier detected CO in exhaled breath of guinea pigs (Phillips, 2002). Since then, colorimetric assays and gas chromatography columns have been used to detect VOCs (volatile organic compounds) in (human) breath in quantities varying from millimolar (10^{-3}M) to picomolar (10^{-12}M) concentrations. The latter gas sensitivity limit was achieved by Linus Pauling's gas-chromatography-based breath analysis device in 1971. Among the 400 compounds of which the human breath consists, only 30 have been

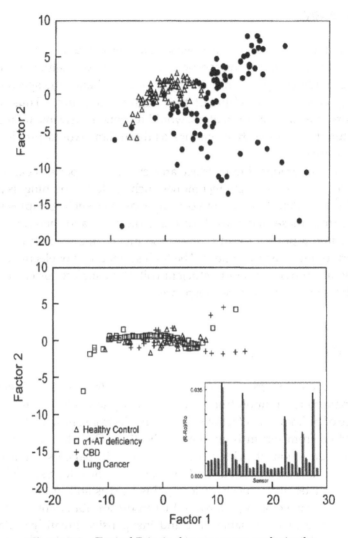

Figure 5.6. Typical Principal components analysis plot.

identified and most of them are potential indicators (markers) of more than one type of diseases. Breath VOCs, in particular, are providing new markers of oxidative stress condition (Kharitonov and Barnes, 2001). Oxidative stress refers to "an imbalance between the production of various reactive species affecting the ability of the organism's natural protective mechanisms to cope with these reactive compounds and prevent adverse effects" (quote from Brunswick Laboratories). The following effects result: peroxidation of lipids, oxidative attack on proteins and oxidative damage to DNA. In order to assess the degree of oxidative damage on lipids, proteins and DNA (i.e. the oxidation stress status), specific biomarkers need to be detected and monitored.

Volatile organic compounds in exhaled breath may be used to study the mechanisms of human metabolism fast and efficiently, thus enabling the early identification of diseases (such as asthma, SARS, or lung cancer, which cause oxidative stress). Currently, there are no direct measures for these diseases in clinical practice, only invasive procedures (e.g. fiber optic bronchial biopsies) (Smith *et al.*, 2004). Non-invasive monitoring may assist in differential diagnosis of pulmonary diseases, assessment of disease severity and response to treatment.

Medical studies reported recently have associated certain gaseous constituents of the human breath with specific types of diseases, and have addressed the importance of diet, dental pathology, smoking, etc. on determining the physiological levels of the marker concentrations in exhaled breath (Risby and Sehnert, 1999). Inflammation and oxidative stress in the lungs can by monitored by measuring the changes in the concentration of the following gases (Studer *et al.*, 2001): NO (which has been widely studied as a bio-marker), and it's related products NO_2-(nitrite) NO_3- (nitrate); exhaled CO (also a marker for cardiovascular diseases, diabetes, nephritis, bilirubin production); exhaled hydrocarbons of low molecular mass, such as ethane, n-pentane; ethylene, isoprene (hydrocarbon affected by diet which is a marker for blood cholesterol levels) (Karl *et al.*, 2001); acetone; formaldehyde; ethanol; hydrogen sulfide, carbonyl sulfides, and ammonia/amines. For example, measurements of exhaled ammonia may differentiate between viral and bacterial infections in lung diseases to justify the use of antibiotics.

For quantitative medical diagnostics, more that one type of gases should be monitored simultaneously, as markers are affected differently in different diseases. E.g. consider the reported studies focusing on NO, CO, and 8-isoprostane. The following trends in the changes of these markers are associated with each disease: asthma, chronic obstructive pulmonary disease (COPD) /chronic bronchitis, cystic fibrosis (CF), shown in the Table 5.1 below (from Kharitonov and Barnes, 2001):

Furthermore, the specificity of a given exhaled biomarker for the diagnosis of a (lung) disease (such as acute allograft rejection in lung transplant patients) may be improved by including measurements of additional compounds in breath analysis (such as S-containing compounds) (Studer *et al.*, 2001). Therefore, for precise diagnosis, an array of a small number of sensors that are selective to specific biomarkers is required.

Identifying these signaling metabolites (disease markers) and measuring them in trace concentrations is not a trivial problem. The low concentrations (ppb) of analyte molecules required (see Table 5.2) is a major challenge, along with the specificity to a given analyte. On the other hand, the benefit of developing this technology is tremendous. Early detection and effective treatment of cancer is a life-saving practice that nanomateials-based sensors may soon enable.

CASE STUDY: Selective, handheld, acetone detector for monitoring diabetes

Among the hundreds of VOCs present in human breath, acetone has been identified to be a biomarker of diabetes, esp. type-1 diabetes. To measure acetone selectively, ferroelectric WO_3 (ε-WO_3) nanoparticles were chosen because of their affinity to acetone. Using a rapid solidification technique, the acentric ε-WO_3 phase, which usually exists below -40°C, was captured as a metastable phase. Chromium dopants create Cr=O terminal bonds outside the matrix and form a

Table 5.1 Observed trends in the changes of 3 disease markers in human breath (the number of symbols used indicates the intensity of measured changed in the concentration of the metabolite in exhaled breath.

	NO	CO	Isoprene
Asthma	↑↑↑	↑	↑↑
COPD	↑	↑↑	↑↑
CF	↓	↑↑	↑↑

Table 5.2 Disease markers related to oxidative stress and their respective physiological concentration ranges in the human breath as measured by other workers.

Gases	Physiological Ranges In Human Breath
Hydrocarbons	
Ethane	1–11 ppb
Pentane	less than ethane
8-Isoprene	55–121 ppb; 12–580 ppb; 50–1000 ppb
Other gases	
Methanol	160–2,000 ppb
CO	< 6ppm
NO	1–9 ppb lower respiratory tract; 200–1000 ppb upper respiratory tract; 1000 – 30000 ppb nasal level
Acetone	293–870 ppb; 1.2–1,880 ppb
Ethanol	27–153 ppb; 13–1,000 ppb
Ammonia / (Urea)	422–2389 ppb; 200–1750 ppb

chromate layer on the surface of each WO_3 nanoparticle (Wang, 2008). This layer prevents particles from changing size or structure. In this way, the ε-WO_3 phase is able to exist above RT or even higher. This is the first time anyone has reported obtaining and maintaining this polymorph above RT. The mechanism controlling the material's specificity to acetone appears to be the following: that the dipole moment of a polar molecule may interact with the electric polarization of some ferroelectric domains on the surface. This interaction would then increase the strength of molecular adsorption on the material surface. The ε-WO_3 is a type of ferroelectric material with a spontaneous electric dipole moment. The polarity comes from the displacement of tungsten atoms from the center of each $[WO_6]$ octahedra. On the other hand, acetone has a much larger dipole moment than any other gas commonly existing in human breath. As a consequence, the interaction between the ε-WO_3 surface dipole and acetone molecules could be much stronger than any other gas, leading to selective acetone detection.

The detection requirements of acetone in human breath are concentrations of < 0.8 ppm for a healthy person and > 1.8 ppm for a diabetic patient. Introducing repeated cycles of acetone gas flow, the sensitivity did not change indicating good stability of the sensor. In addition, the sensor responds to acetone exposure very quickly, in less than 10 seconds. Therefore, the ε-WO_3 nanoparticle-based sensor is capable of real-time, fast-response, stable and highly sensitive detection of acetone gas. Considering that there are hundreds of gases in human breath, the selectivity of such a sensor to acetone is very important. The relative response of this sensor to a series of interfering gases, including CO, NH_3, methanol, ethanol, CO_2, NO and NO_2, which are commonly present in human breath was also tested. Compared to acetone, the sensor shows substantially lower sensitivity to other gases. Ethanol and methanol also belong to VOCs. However, the sensor sensitivities to 1 ppm of these gases are even lower than that of 0.2 ppm of acetone. More interestingly, although pure ε-WO_3 is generally considered to be very sensitive to NOx gases, the present flame-made WO_3 sensor shows quite low sensitivities to NO and NO_2, less than 1.1 when the gas concentrations are 1 ppm. Finally, this sensor shows a very low response to NH_3 and almost no response to CO.

The sensor has a very good selectivity to acetone at 400°. To make the portable breath analyzer, the sensor was attached onto a heater and then embedded into a circuit board. Fig. 5.7 also shows the photograph of the prototype. The bottom-left part is the sensor. It will be isolated from the environment by a chamber and specially designed channel allows human breath flow or controlled gas flow to go through the chamber and interact with the sensor. Although this device has not yet been utilized to analyze exhaled human breath, it is expected to serve as excellent acetone detector in human breath and finally a promising tool for non-invasive diabetes diagnostics.

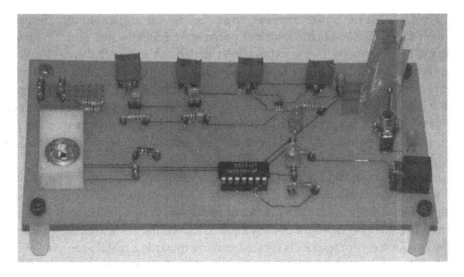

Figure 5.7. Acetone breathalyzer prototype. For color reference, turn to page 153.

5.4 TISSUE ENGINEERING

Tissue engineering aims at restoring, maintaining or improving tissue function (Badylak, 2002). This is occurring in nature using the 3D-structure of Extra-Cellular Matrix (ECM)- the natural scaffold- that allows cells to grow, proliferate and differentiate within it (Griffith, 2002; Kim and Mooney, 2001; Berthiaume *et al.*, 1996). ECM is a complex, three-dimensional ultrastructure of proteins, proteoglycans and glycoproteins, used to promote cell growth in native tissue (Cooper, 2000). In fact, there are many different types of ECMs for different parts of the body. For example, fibrous proteins are dominant material in tendon; polysaccharides are found largely existing in cartilage and so the forth. ECM provides attachment sites and mechanical support for cells. In addition to the bioactive functional motifs present, the topology of ECM has also been found to affect the cell structure, functionality and its physiological responsiveness. The geometry of the natural matrix was reported to modulate the cell polarity. Thyroid cells, smooth muscle cell and hepatocytes are different types of cells found to be affected by ECM's topology, with 3D-structures inducing cell differentiation more effectively than 2D configurations. The arrangement of ECM's configuration involves multiple length scales, layers and morphologies.

The extracellular matrix (ECM) isolated from the porcine small intestine or urinary bladder has been shown to provide an environment suitable for regenerating many different types of tissue including small diameter arterial grafts (Antucci and Barber, 2005), vena cava (Hodde *et al.*, 2002), urinary bladder (Colvert *et al.*, 2002; Knapp *et al.*, 1994), all suggesting its potential as a scaffold capable of tissue remodeling. No immune response has been noted in over 150,000 patients in whom the extracellular matrix isolated from the porcine small intestine has been implanted (Badylak, 2004). Still, there are limitations in the reproducibility of the topology of the material, sterilization, etc.

The ideal tissue repair material, thus, one might argue, should consist of syntheti biomaterials, such as natural polymers mimicking the mechanical and biological functionality of the extracellular matrix. A major challenge for tissue engineering researchers is to find materials and processing techniques that allow them to produce ECM mimicking scaffolds that promote cell growth and organization into a specific architecture, inducing cell differentiation and subsequent cell function. Electrospinning is considered as a potential single-step method for producing three-dimensional scaffolds of tailored topological design mimicking the morphology of extracellular matrix.

Electrospinning can generate polymeric fibers with diameter ranging from a few nanometers to several tens of micrometers and structures with open porosity. By applying an electric field to a polymeric melt or solution at sufficiently high concentration contained in a syringe, forcing it to exceed the surface tension at the tip of the needle, a continuous jet is formed moving towards a metal grounded collector placed not far away resulting in continuous fiber formation in non-woven mats. By varying the polymer concentration (surface tension) and processing parameters including the strength of electric field, tip-target distance and the flow rate of polymeric solution or melt, the morphology of the mats can be easily modified in a continuous manner, as described in the case studies below.

5.4.1 Electrospun Nanocomposite Mats for Tissue Engineering

The biomedical role of nanofibers has extended into various specific applications with tissue engineering being the most widely studied. Biodegradable polymers can be directly electrospun onto an injured location of skin to form a fibrous mat dressing. Wounds covered with polymer membranes, that encourage formation of normal skin growth and eliminate formation of scar tissue that would occur in a traditional treatment, heal faster (Fertala *et al.*, 2001). Scaffolds and synthetic matrices mimic the structure and biological functions of the natural extra cellular matrix (ECM). Nanoscale fibrous scaffolds provide an optimal template for cells to seed, migrate and grow (Huang, 2000). Electrospun biocompatible polymer nanofibers can also be deposited as thin porous mats onto a hard tissue prosthetic device designed to be implanted into the human body. Membranes served as an inter phase between the prosthetic device and the host tissues. This reduces the stiffness mismatch at the tissue-device interface and hence prevents the device failure or infection after the implantation (Buschko, 1999).

The introduction of composite materials into the electrospun mats can amplify their functionality by strategic incorporation of specific species. A candidate material for tissue engineering scaffolds, wound dressings, or homeostatic products was developed by electrospinning fibrinogen [48]. Natural silk fibers of Bombyx mori and Nephila clavipes with diameters orders of magnitude smaller than natural silk spun by most silkworms and spiders were prepared by electrospinning (Zarkoob *et al.*, 2004). Annealing process proved the nanofibers have a crystallographic order similar to that observed in the original natural fibers. The availability of nanoscale silk fibers can introduce a set of new possible uses at a scale not explored before.

Another group has electrospun a mixture of aqueous regenerated silkworm solution with polyethelyne oxide (PEO) (Jin *et al.*, 2004). The silk fibroin-based fibers with an average diameter of 700 nm were tested for supported human bone marrow stromal cell attachment and proliferation over 14 days. Polymer supplied good mechanical properties to the electrospun mats, and 1–2 days following PEO extraction, those effects were abolished and proliferation was observed. After 14 days of incubation, the electrospun silk mats supported extensive human bone marrow stromal cells proliferation and matrix coverage.

Bioabsorbable polymer, Polydioxanone (PDS), first developed specifically for wound closure sutures, was electrospun and formed fibers with diameters ranging from about 0.18 to 1.4 μm as a direct function of the concentration of the solution (Boland, 2005). Mechanical testing revealed anisotropic properties of the oriented mats deposited on a rotating collection plate. New types of guided bone regeneration membranes were synthesized by electrospinning polycaprolactone (PCL) and PCL/CaCO$_3$ composite nanofibers (Fujihara *et al.*, 2005). The authors concluded that membranes rich in PCL had better cell attachment and proliferation than those of $CaCO_3$ rich membranes.

Tissue engineering was advanced when synthetic polymer/DNA composite scaffolds for therapeutic application in gene delivery were electrospun composed of poly(lactide-co-glycolide) (PLGA) and poly(D,L-lactide)-poly(ethylene glycol) (PLA-PEG) (Liu *et al.*, 2003). Variations in the PLGA to PLA-PEG ratio were observed to vastly affect the overall structural morphology, rate and efficiency of DNA release from 68–80% of the initially load. The synthetic tensile moduli and strain resemble those of skin and cartilage.

Cellulose acetate (CA) thin, porous membranes were produced by electrospinning precursor polymer solutions in acetone at room temperature. These membranes were used as scaffolds for microvascular cells growth (Han *et al.*, 2005). Human umbilical vein endothelial cells were obtained as first passage cultures. Cells were incubated at 37° in a 5% CO_2 humidified atmosphere. The electrospun materials were tested for their effect on cellular viability. Cells were trypsin digested to transfer onto $1cm^2$ area membrane at a dilution of 1:1 or 1:2. On days 3–5 post seeding, cellular viability was assessed. Living endothelial cell in culture were viewed with fluorescence and phase contrast microscopy. The structure of the membranes that were produced mimic the topology and porosity of extracellular matrix (ECM) in two key ways: (i) the fiber diameter resembled the extracellular protein fiber diameter, thus enabling cellular attachment and facilitating cellular migration; and (ii) the porosity simulateds that of extracellular matrix such that microvascular capillary tube formation *was* enhanced (Han *et al.*, 2005).

5.4.1.1 Biopolymers as electroactive scaffold components

The use of electroactive polymers in cardiac tissue engineering has only recently been explored to facilitate the culture of electrically excitable cells. Polyaniline (PANI), is such a polymer formed from the oxidation polymerization of the aniline monomer. The base structure of PANI, shown below in Fig. 5.8, consists of a reduced unit or benzenoid attached to an amine (-NH-), and an oxidized unit or quinoid attached to an imine (-N=). The amines and imines are the nitrogenous centers of polyaniline and the ratio of amines to imines dictates the oxidation state. Moreover, the imine sites act as centers for electroactive reactions.

The adhesion and proliferation of H9c2 rat cardiac myoblasts on thin films of polyaniline (PANI) showed that it is a biocompatible conductive polymer that offers the capability of continuous electrical stimulation for up to 100 hr in culture. More recently, the same group prepared electrospun PANI-gelatin composite fiber mats (PANI concentration was kept low (<3% w/w)) and tested the mats as substrates for supporting the same cell growth. The findings suggested that cells preferentially grow and spread on smaller fibers, as these resemble their *in-vivo* ECM. Thus, electrospun fiber mats that mimic the ECM architecture and which contain conductive polymer components have emerged as promising candidates for functional, "active", cardiac tissue constructs.

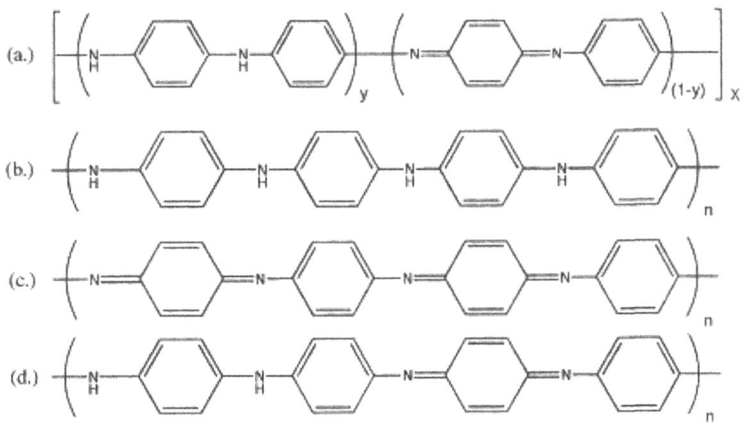

Figure 5.8. (a) Base structure of Polyaniline (b) for $y = 1$ the oxidation state is leucoemeraldine, (c) for $y = 0$ the polymer is in the pernigraniline oxidation state and (d) for $y = 0.5$ the polymer is in the emeraldine oxidation state.

Table 5.3 Dimension detail of model ECM.

	Fibrous Layer	Cellular Layer	Dense Layer
Dimension	Fiber Diameter $1.7 \pm 0.1\,\mu$m	Average pore size: $36.6\,\mu$m Average pore height $16.4 \pm 7\,\mu$m Average pore length $68.3 \pm 5\,\mu$m	Thickness: $2.0\,\mu$m

CASE STUDY: ECM-mimicking 3D scaffolds

One type of ECM obtained from porcine urinary bladder matrix (UBM) has found potential use as a keratinocyte growth substrate (Badylak, 2002). Figure 5.9 shows a scanning electron micrograph of the cross sectional view of the UBM specimen. A dense basal layer, followed by a multilayered (3-layer) cellular-type structure with flattened, elongated, ellipsoidal-shaped pockets, topped by a non-uniformly, loosely shaped fibrous layer are observed. Table 5.3 summarizes the dimensions of the key features of the UBM's architecture. The structure of the fibrous layer consists of uniform size fibers with an average diameter of $1.7 \pm 0.1\,\mu$m. The cellular-type layer's vary widely pores with the smallest being $11.8\,\mu$m and the largest $72\,\mu$m (average pore size estimated to be $36.6\,\mu$m). The dense bottom level has a thickness of $2.0\,\mu$m. This structure is what we will tried to mimic by fabricating cellulose acetate structures with the same shape and appearance.

5.4.1.2 Cellulose Acetate Electrospun Scaffolds

As the most plentiful organic compound existing in the world, cellulose is the primary component of the higher plants' cell walls. The structure of cellulose can be identified as a long chain polysaccharide. CA is one of the cellulose derivatives.

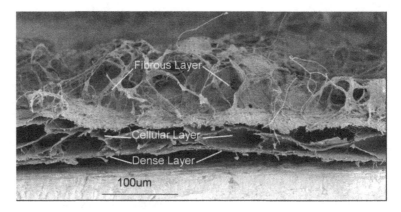

Figure 5.9. Cross-sectional view of the UBM specimen.

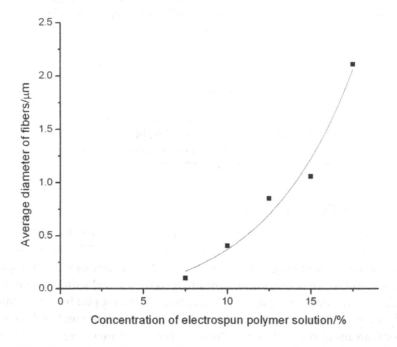

Figure 5.10. Dependence of average diameter vs. concentration of polymer solution.

It can be dissolved in organic solvents such as acetone and acetic acid. CA has very low water solubility that is an advantage when using it for scaffold fabrication (Han, 2007; Entcheva *et al.*, 2004). In the presence of adhesion proteins, such as fibronectin, CA scaffolds used for cardiac cell growth were found to boost the cell growth and helped increase their connectivity. The cell adhesion properties were better than that of other polymeric artificial scaffolds. CA has also been reported to have good biocompatibility (Demir *et al.*, 2002). CA scaffolds are prepared by electrospinning in this work.

The measured effect of the process parameters on the fiber structures is summarized in Figs. 5.10–5.12. When the polymer entanglement is not sufficient resulting in the instability of the polymer solution jet, polymer fibers with beads can be observed. Solution viscosity and net charge density are two major factors for forming beads (Fong *et al.*, 1999). The largest fiber diameters can be obtained for high concentrations of the polymer solution (17.5% w/v), high flow rate (160 µl/min) and low strength of electric field (7 kV). At very low solution concentrations, 7.5 % w/v, droplets of polymers were generated and beads were seen under high magnification of SEM. The porosity of such obtained mats is very low and dense. When the concentration went up to 17.5% w/v, the morphology of the mats was open rather than compact as observed for the low concentration polymer solution. Although mechanical tests were not performed on the respective structures, the strengths of two different mats are apparently different: dense beads-like mats are much stronger than loose fiber mats.

In order to produce the structure of UBM scaffolds, different CA solutions and processing parameters were designed, according to the knowledge obtained from the preliminary experiments with the single layer mats. In order to mimic the bottom layer of ECM, the structure of the mat should be very dense and flat. From the previous experience, low concentration of polymer solution, low flow rate and high electric field will result in dense structure of mat. The solution with concentration of 7.5% w/v was first used as electrospun material to produce the bottom layer one under 30 µl/min and 19 kV.

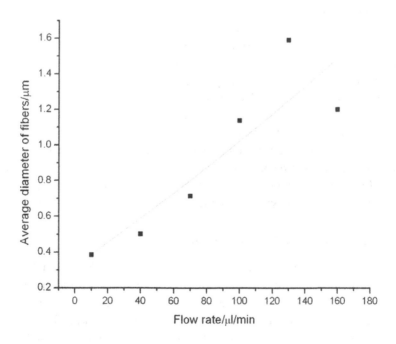

Figure 5.11. Dependence of average diameter vs. flow rate.

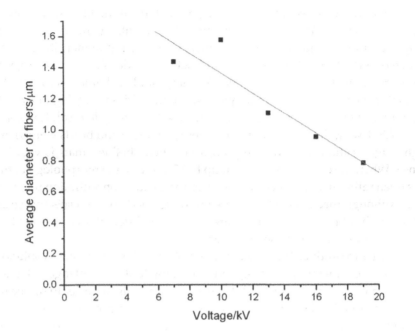

Figure 5.12. Dependence of average diameter vs. electric field.

Considering the middle layers and the bottom layers together, it appeared that the two layers were connected at regularly spaced intervals, separated by large pores in between. Therefore, by mimicking this, the structure of the middle layer needs to be designed between that of a dense and a loose structure. Before electrospinning the middle layer, a copper wire pattern was created to introduce porosity between the first and second layers. Copper wire coils with a diameter of 300 μm were used to make a digitated-shaped template with a separation distance of 1 mm between two adjacent coils. This template served as a physical separator to induce the channeled (cellular-like) configuration. This "removable" pattern did not interrupt the continuity of the electrospinning process.

The solution at concentration of 10.0% w/v was then electrospun onto the initial layer under the same conditions. The top layer of ECM consists of randomly oriented fibers with large porosity. The processing condition selected for the top layer were 17.5% w/v polymer concentration under 160 μl/min and 7 kV.

5.4.1.3 Morphology of the bio-Mimicking Scaffolds

Figure 5.13 shows the SEM image of this designed UBM- mimicking scaffold. Although only one middle layer is shown here, the structural features are consistent with the UBM's. The size of the structure of the fibrous layer: average diameter of fibers is 2.0±0.5μm; the cellular layer includes ellipsoidal-shaped pores with average pore size 357.2μm (closely related to the copper coil size used for patterning purposes), the dense layer's morphology is identical to that of the respective layer of the UBM. It has been found in studies with ECM scaffolds and hepatocytes

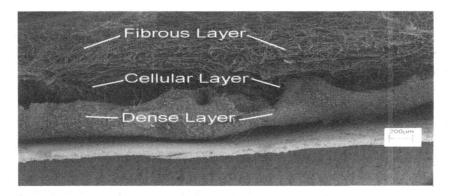

Figure 5.13. SEM image of multilayer structure scaffold.

that sandwiched configurations (like the cellular one achieved herein) reduce cell spreading and enhance the expression of differentiated cell function (Berthiaume *et al.*, 1996). Therefore, it makes them very promising for the next generation of artificial scaffolds for tissue and organ growth. The fact that such structures may be fabricated in a single step process using minimal patterning that does not disrupt the manufacturing process is an innovation that may prove useful for scaffold fabrication for tissue engineering applications. This is the first reported effort of using a single technique to control both the scaffolds architecture and chemistry.

CASE STUDY: Nanocomposite Scaffolds for Bone Engineering

Bone tissue engineering has emerged as a subject of significant importance in the past decade as most of the serious injuries related to bone are still unrecoverable and changes in the bone structure due to an injury dramatically alters ones' body equilibrium (Salgado *et al.*, 2004). There are about 1 million cases of skeletal defects a year that require bone-graft procedures for reconstruction (Yaszmeski *et al.*, 2000). Cell-based approach for bone grafting that makes use of scaffolds fabricated from synthetic biomaterials such as metals, ceramics, polymers and composites has been recognized as a promising technique and is used to overcome the drawbacks of autologous and allogenic bone-graft techniques (Meijer *et al.*, 2007).

The requirements for a successful scaffold for bone tissue engineering are biocompatibility, osteoinductive properties and high porosity with interconnected pores (van Lenthe *et al.*, 2007). Fibrous scaffolds are being widely used in tissue engineering as they have interconnected pores through out the matrix that enables the cells to grow into the matrix. Porosity is a very important property of the scaffolds because of the existence of the need for three-dimensional cell and bone growth (Salgado *et al.*, 2002). Pores in a tissue-engineering scaffold make up the space in which cells reside (Yoshimoto *et al.*, 2003). Pore properties such as shape, size and volume are thus key parameters that determine the usefulness of a scaffold. High porosities have been reported to provide more structural space for cell accommodation. For bone regeneration, pore sizes between 100 and 350

μm and porosities of more than 90% are preferred (Hunter *et al.*, 1995). A stiffness value of -0.41 N/m has been reported for electrospun CA scaffolds with a polymer concentration of 16.5% (Rubinstein, 2007). This suggests that the electrospun fibrous CA scaffolds used in this study are very flexible in nature and the pores can dynamically expand to accommodate cell growth into the scaffold.

Hydroxyapatite (HA), a bioactive material known to promote differentiation of osteoblastic cells *in vitro* (Marques and Reis, 2005) has been used in bone tissue engineering due to its biocompatibility and osteoinductive/osteoconductive properties. Nanocrystalline HA crystals are similar to bone apatites and are known for their excellent mechanical properties and bioactivity (Zhang *et al.*, 2006). Human osteoblasts, bone forming cells derived from stem cells of the mesenchymal lineage (El-Amin *et al.*, 2006), are typically used to evaluate the potential of a material as bone scaffold by carrying out cell attachment and proliferation studies. The aim of this case study was to assess the potential of cellulose acetate and its hybrid with nanoscale hydroxyapatite scaffolds fabricated by electrospinning process for osteoblast attachment and proliferation.

SEM micrographs of electrospun scaffolds revealed that the CA fibers were found to be ribbon shaped. The average width of the ribbons was $3.1 \pm 0.5\,\mu$m. Their thinner section was $0.8 \pm 0.3\,\mu$m. The CA-HA hybrid mats consist of fibers with wire-like morphology, of an average diameter of $440 \pm 170\,$nm. The average size of hydroxyapatite nanoclusters in the matrix is $970 \pm 650\,$nm. The pore diameter in the CA scaffold varied between 8–12 μm with a porosity of around 30% and the pore diameter in CA-HA varies between 3–7 μm with porosity relatively less than on CA scaffolds. This study explores the feasibility of CA and CA-HA nanocomposite hybrid mats as potential scaffolds for bone tissue engineering.

The MTS assay suggests a significant increase in the number of cells between day 1 and day 3 on both the CA and the CA-HA hybrid scaffolds. The PicoGreen assay, a quantitative measure of double stranded DNA, is an alternative measure of cell proliferation. This assay also indicated an increase in the number of proliferating cells between day 1 and day 3 for both scaffolds. Thus, it is clear that these scaffolds support cell proliferation.

The osteoblasts seeded on the CA scaffold were attached along the thinner section of the ribbon-like CA fibers. The cells showed a rounded morphology. The average diameter of the cells was $8.5 \pm 1.4\mu$m. The cells were typically found attached on a single fiber. Figure 5.14 shows scanning electron micrographs of a cell attached to a single CA fiber.

The osteoblasts seeded on the CA-HA composite scaffolds were found attached to several fibers forming an interconnected network. There are only a few cells with rounded morphology. The average diameter of the rounded cells was 6.88 ± 1.53 μm. The osteoblasts seeded on the CA-HA scaffold however showed mainly a flat morphology. Most of the cells appeared spread out. Fig. 5.15 shows typical SEM images of flat cell spreading, taken at different magnification and using different voltage.

Thus, the structural characterization studies carried out using electron microscopy suggests that the osteoblasts prefer to attach to thinner fibers. It was

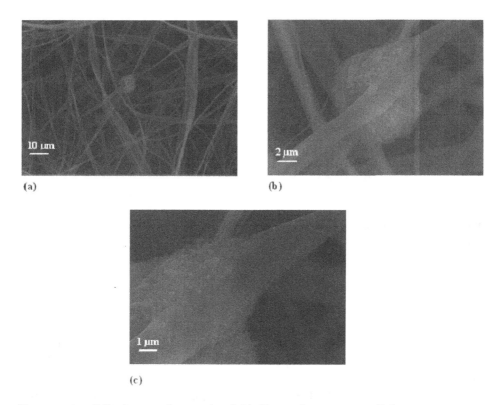

(a)   (b)

(c)

Figure 5.14. Cells that attach strongly of CA fibers releasing extracellular matrix components.

evident from the fact that the cells seeded on CA scaffolds adhered along the thinner section of the ribbon-like fibers. Cells seeded on CA-HA fibers were observed to attach to multiple fibers elaborating a large number of anchoring ligands. This is related to the fact that the fiber diameter of CA-HA fibers was significantly smaller compared to the CA fibers, thus having a direct influence on the cell attachment behavior. It was clear that hydroxyapatite nanoclusters provided anchoring sites for the cells in the CA-HA mats. The cell density was also observed to be higher in the case of CA-HA hybrid mats.

Osteoblasts are anchorage-dependent cells (Yoshimoto *et al.*, 2003, hence they depend on the high surface area of the scaffolds and the porosity for the attachment and migration of cells in the scaffold (Yoshimoto *et al.*, 2003). The electrospun CA and CA-HA hybrid scaffolds have fine fiber diameters indicating a high surface area-to-volume ratio suitable for osteoblast attachment and migration. Yoshimoto *et al.*, 2003 have electrospun microporous, non-woven PCL scaffolds with three-dimensional fibrous mesh consisting of randomly oriented fibers with diameters ranging from 20 nm to 5μm for bone tissue engineering. Mesenchymal stem cell-derived osteoblasts seeded on the scaffolds have been reported to migrate into the scaffold and produce an extracellular matrix of collagen throughout the scaffold, which was confirmed by histology and immunohistochemistry studies. This is

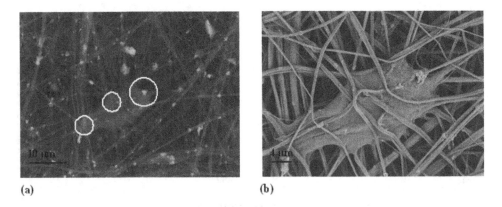

(a)   (b)

Figure 5.15. Cells anchor on nanocrystalline HA aggregates and spread out, promoting differentiation to osteoblasts.

relevant to this case study, as observed in Fig. 5.14, a prominent step-wise morphology of extracellular matrix agents (possibly type I collagen) secreted by the cells is shown.

One of the most important aspects of a successful scaffold is its capacity for cell attachment and interaction. It has been shown that the degree of cell attachment has a direct influence on cell motility, proliferation rate and control of phenotype. Hunter *et al.*, 1995 have showed that one of the main regulators of proliferative rate in anchorage dependent cells is their shape. Cells in a rounded configuration are reported to divide at lower rates than those flattened and well spread. Cell shape has been shown to have a direct influence on DNA synthesis and growth in non-transformed cells. A decrease in DNA synthesis with the change of cell shape towards a spheroidal conformation has also been reported. It is thus understood that a critical cell shape is required in anchorage-dependent cells for controlled growth (Folkman and Moscona, 1978). In this study, the $SaOS_2$ cells attached on CA fibers show rounded morphology and do not show any spreading, suggesting that these cells attached on CA may divide at a lower rate. It has also been reported by Hunter that a biomaterial that does not allow spreading might favor a change of cell phenotype from osteoblast to fibroblast. Hydroxyapatite has been reported to increase the amount of cell spreading when incorporated in starch based polymer scaffolds (Marques and Reis, 2005). This is in agreement with the results obtained in this study. The cells attached to CA-HA hybrid fibers appear flattened and well spread out, and considering relevant literature, it is understood that these cells might synthesize more DNA and divide at a higher rate. This suggests that the hydroxyapatite nanoaggregates play an important role in cell spreading and differentiation. *It was observed that the composition and topography of CA-HA nanocomposites has an influence on the morphology of cells.*

5.5 NANO-ELECTRO-CHEMO-ACTUATORS

5.5.1 *Polyaniline Actuation Principle*

The actuation of PANI films is due to dimensional changes occurring upon a reversible redox reaction. As the polymer becomes protonated (it's been added extra H+ ions) it increases in size, and in an alkaline solution de-protonation occurs and the mat should decrease it's size; this is a pH dependent redox reaction thus PANI may be used as an actuator in the presence of different pH media (Gao *et al.*, 2003). PANI can also actuate in the presence of volatile gasses (chemoactuation) or in the presence of an electric field.

5.5.2 *Polyaniline Hybrid mats*

Preliminary studies by the author's group on PANI hybrid mats were carried out to access their potential for actuation. Solutions of 0.1 g of LEB PANI in 10 ml of HCl were kept in a glass vial over night. The solution was subsequently centrifuged and the acid was removed. To the remaining precipitate, water was added. The new solution was centrifuged again and this procedure was repeated 6 times. The acid-treated PANI was then dried at room temperature and it was then added to acetone. The mixture was sonicated for 1 hour, following which, cellulose acetate was added to the acetone/PANI mixture in a ratio of PANI: CA equal to 1:5; the total polymer concentration in acetone was 5%. Thin films of the hybrid were made initially by casting and ambient drying of the PANI/CA solution in acetone.

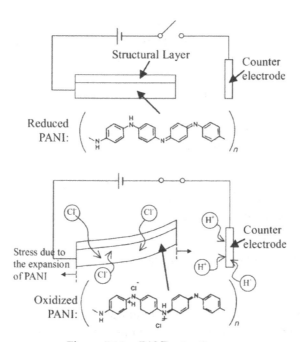

Figure 5.16. PANI actuation.

Figure 5.17. Chemo-mechanical actuation of PANI-CA hybrid films in acetone vapor. For color reference, turn to page 153.

Thin strips with dimensions of 5 mm width and 40 mm length were cut. The figure below shows one such strip held above a beaker containing organic solvents (acetone in this case); acids such as HCl and acetic acid were also used in our work. In all cases, chemical-based actuation (bending) of the strip was observed. After removing the strip from the headspace of the chemical, it returned to its original shape (see Fig. 5.17). This is an example of chemo-mechanical actuation of PANI hybrids. This type of actuation is common in biological muscles.

Several studies have explored the electrical conductivity of PANI-CA hybrids, observing that since CA is hydrophobic, the hybrid can exhibit a percolation threshold of 0.006, which is significantly lower than the other base polymers (e.g ~4% for PEO) thus being able to retain charge. Electrical actuation of these mats and their electrospun counterparts need to be further explored in the future.

5.6 OVERVIEW AND FUTURE TRENDS

The focus of this book has been on the use of nanomaterials and one-dimensional nanostructures, in particular, to functional applications, such as chemical vapor detection for environmental safety as well as personalized medicine diagnostics. The emphasis has been placed on metal oxides, as they are the most selective, sensitive, and reliable chemo-sensing elements known to date, compared to the nanostructured polymers, carbon nanotubes, or microcantilever competitors. New breakthroughs achieved thanks to nanotechnology include metastable polymorph selection via controlled nanoscale synthesis/processing, which made the appropriate crystallographic configuration of a single metal oxide that exhibits specificity to a given gas readily available. This, in turn, led to the development of single breath analysis prototype devices for non-invasive diagnostics for oxidative stress, diabetes, and hopefully ones for cancer detection soon.

Another nanotechnology breakthrough reported in this book involves the single step synthesis of single crystal metal oxide nanowires of "extreme" aspect ratio, by means of a novel nanomanufacturing technique. The preparation of single crystalline nanowires is the most prominent trend for the future of nanostructured materials. A recent review by Shen *et al.*, 2009, highlights the importance of having monocrystalline 1D nanostructures and the need for bottom up synthesis methods, suggesting templating as the most likely approach to be followed to achieve single nanowire devices. Such devices currently exist at the proof of principle stage, including a single nanowire field effect transistor, a transparent active matrix organic light emitting diode display, lasers, and piezoelectric generators. The same review emphasizes the anticipated use of metal oxide nanowires in solar cell and photocatalytic applications for energy generation and environmental remediation, respectively, as well as in nanoemitters for use in displays and nanocharacterization equipment.

Another theme of this book is electrospinning, a unique and versatile nanomanufacturing process used to produce nanofibrous polymers, and their composites, has become already a high profile tool to engineer scaffolds for implants, drug delivery vehicles, and nanoactuators. It has been demonstrated in this book how electrospun nanofibrous scaffolds of biopolymers can be constructed in complex, porous, 3D architectures mimicking the bio-scaffold-the ExtraCellular Matrix (ECM) topography, in a singe step process. This is a breakthrough for tissue engineering nanomanufacturing (see review by Xie *et al.*, 2008). Furthermore, it has been discussed how nanofibers promote cell growth and proliferation. The current trend in tissue and bone engineering is to utilize stem cells grown on well defined nanostructures (see Smith *et al.*, 2008); this is to prove that the nanoscale geometry is key to controlling cell differentiation to a particular phenotype (functionality). The book has also touched upon the topic of conducting polymer nanofibers (polyaniline in particular) for use in tissue engineering and nanomedicine. A couple of recent reviews have summarized the trends in conducting polymer nanowires. Given that these materials may change their conductivity by 15 orders of magnitude via doping and redox reactions, while at the same time they are processed and operate at room temperature, they offer unique properties which makes them attractive as "active" scaffolds. One example is a recent application where anticancer drugs were encapsulated into such electrospun fibers producing promising implants for brain tumor treatment.

Encapsulation was another of the themes of this book, involving either the insertion of biomolecules in electrospun nanofibers or the entrapment of bio-species in metal oxide sol-gels. There are enormous opportunities to exploit these novel nanocomposites for biosensing and controlled drug delivery (theragnostics) applications as well as for energy harvesting and production. Our world has just begun to grasp the importance of nanomaterials and the nanotechnologists have already brought key applications to light. The future of nanomaterials looks bright indeed!

References

[1] Adamson DH, Dabbs DM, Morse DE and Aksay I, "Non-peptide, silicatein a inspired silica condensation catalyst", *Polymeric Materials: Science and Engineering*, (2004), **90**, 239–240.

[2] Affleck R, Haynes CA and Clark DS, "Solvent dielectric effects on protein dynamics", *Proc. Natl. Acad. Sci. USA*, (1992), **89**, 5167–5170.

[3] Al-Saraj M, Abdel-Latif MS, El-Nahal I and Baraka R, "Bioaccumulation of some hazardous metals by sol-gel entrapped microorganisms", *Journal of Non-Crystalline Solids*, (1999), **248**, 137–140.

[4] Angadjivand S, Schwartz MG, Eitzman PD, Jonesm and ME, "Method and apparatus for making a nonwoven fibrous electret web from free-fiber and polar liquid", *US Patent 6*, (2002), **375**, 886.

[5] Ansari A Jones CM, Henry ER, Hofricher J and Eaton WA, "The role of solvent viscosity in the dynamics of protein conformational changes", *Science*, (1992), **256**, 1796–1798.

[6] Ansari, ZA, Ko TG and Oh JH, "CO-sensing properties of In_2O_3-doped SnO_2 thick-film sensors: Effect of doping concentration and grain size", *IEEE Sens. J.* (2005), **5**, 817–824.

[7] Arterbery VE, Pryor WA, Jiang L, Sehnert SS, Foster WM, Abrams RA, Williams JR, Wharam MD, Jr. and Risby TH, "Breath ethane generation during clinical total body irradiation as a marker of oxygen-free-radical-mediated lipid peroxidation: A case study", *Free Radical Biology & Medicine*, (1994), **17**, 569–576.

[8] Aslam M, Mulla IS and Vijayamohanan K, "Insulator to metal transition in coulomb blockade nanostrcutures", *Applied Physics Letters*, (2001), **79**, 689–691.

[9] Athreya, SA and Martin DC, "Impedance spectroscopy of protein polymer modified silicon micromachined probes", *Sensors and Actuators a-Physical*, (1999), **72**, 203–216.

[10] Aughenbaugh W, Radin S and Ducheyne P, "Silica sol-gel for the controlled release of antibiotics- II: The effect of synthesis parameters on the *in vitro* release kinetics of vanomycin", *J. Biomed. Mater. Res.*, (2001), **57**, 321–326.

[11] Avnir D, Coradin T, Lev O and Livage J, "Recent bio applications of sol-gel materials", *J. Mater. Chem.*, (2006), **16**, 1013–1030.

[12] Badjic JD and Kostic NM. "Effects of encapsulation in sol-gel silica glass on esterase activity, conformational stability, and unfolding of bovine carbonic anhydrase II", *Chemistry of Materials*, (1999), **11**, 3671–3679.

[13] Badjic JD and Kostic NM, "Behavior of organic compounds confined in monoliths of sol-gel silica glass. effects of guest-host hydrogen bonding on uptake, release, and isomerization of the guest compounds", *J. Mater. Chem.*, (2001), **11**, 408–418.

[14] Ballantine DS, White RM, Martin SJ, Ricco AJ, Zellers ET, Frye GC and Wohltjen H, *Acoustic Wave Sensors*, Academic Press, San Diego, CA, 1997.

[15] Banerjee AN and Chattopadhyay PK, "Recent developments in the emerging field of crystalline p-type transparent conducting oxide thin films", *Prog. Cryst. Growth Charact. Mater.*, (2005), **50**, 52–105.

[16] Bansal V, Rautaray D , Ahmad A and Sastry M, "Biosynthesis of zirconia nanoparticles using the fungus fusarium oxysporum", *J. Mater. Chem.*, (2004), **14**, 3303–3333.

[17] Bansal V, Rautaray D, Bharde A, Ahire K, Sanyal A, Ahmad A and Sastry M, "Fungus-mediated biosynthesis of silica and titania particles", *J. Mater. Chem.*, (2005), **15**, 2583–2589.

[18] Barsan N and Weimar U, "Conduction model of metal oxide gas sensors", *J. of Electroceramics*, (2001), **7**, 143–167.

[19] Barsan N, "Conduction models in gas-sensing SnO2 layers: Grain-size effects and ambient atmosphere influence", *Sensors and Actuators B*, (1994), **17**, 241–246.

[20] Baughman, RH, Zakhidov AA and de Heer WA, "Carbon nanotubes — the route toward applications" *Science*, (2002), **297**, 787–792.

[21] Beece D, Eisenstein L, Frauenfelder H, Good D, Marden MC, Reinisch L, Reynolds AS, Sorensen LB and Yue KT, "Solvent viscosity and protein dynamics", *Biochemistry*, (1980), **19**, 5147–5156.

[22] Belcher AM, Hansma PK, Stucky GD and Morse DE, "First steps in harnessing the potential of biomineralization as a route to new high performance composite materials", *Acta Mater.*, (1998), **46**, 733–736.

[23] Bellingham JR, Mackenzie AP and Phillips WA, "Precise measurements of oxygen-content — Oxygen vacancies in transparent conducting indium oxide films", *Appl. Phys. Lett.*, (1991), **58**, 2506–2508.

[24] Bellissent-Funel MC, "Status of experiments probing the dynamics of water in confinement", *Eur. Phys. J. E.*, (2003), **12**, 83–92.

[25] Berthiaume F, Moghe P, Toner M and Yarmush M, "Effect of extracellular matrix topology on cell structure, function, and physiological responsiveness: Hepatocytes cultured in a sandwich configuration", *Faseb Journal*, (1996), **10**, 1471–1484.

[26] Besanger TR, Chen Y, Deisingh AK, Hodgson R, Jin W, Mayer S, Brook MA and Brennan JD, "Screening of inhibitors using enzymes entrapped in sol-gel-derived materials", *Anal. Chem.*, (2003), **75**, 2382–2391.

[27] Bhatia RB and Brinker CJ, "Aqueous sol-gel process for protein encapsulation", *Chem. Mater.*, (2000), **12**, 2434–2441.

[28] Bidez III PR, Li S, MacDiarmid AG, Venancio EC, Wei Y and Lelkes PI, "Polyaniline, an electroactive polymer, supports adhesion and proliferation of cardiac myoblasts", *J. Biomater. Sci. Polymer Edn.*, (2006), **17**, 199–212.

[29] Bishop A, "Electrospun conducting polymer composites for chemo-resistive environmental and health monitoring applications", *Ph.D.* (2008), SUNY Stony Brook.

[30] Blyth DJ, Poynter SJ and Russel DA, "Calcium biosensing with a sol-gel immobilized photoprotein", *Analyst*, (1996), **121**, 1975–1978.

[31] Bognitzki, M, *et al.*, "Polymer, metal, and hybrid nano- and mesotubes by coating degradable polymer template fibers", *Advanced Materials*, (2000), **12**, 637–640.

[32] Boland, ED, *et al.*, "Electrospinning polydioxanone for biomedical applications", *Acta Biomaterialia*, (2005), **1**, 115–123.

[33] Boninsegna S, Bosetti P, Carturan G, Dellagiacoma G, Dal Monte R and Rossi M, "Encapsulation of individual pancreatic islets by sol-/gel SiO2: A novel procedure for perspective cellular grafts", *Journal of Biotechnology*, (2003), **100**. 277–286.

[34] Böttcher H, Soltmann U, Mertig M and Pompe W, "Biocers: Ceramics with incorporated microorganism for biocatalytic, biosorptive and functional materials development", *J. Mater. Chem.*, (2004), **14**, 2176–2188.

[35] Branyik T and Kunkova G, "Encapsulation of microbial cells into silica gel", *Journal of Sol-Gel Science and Technology*, (1998), **13**, 283–287.

[36] Braun JH, "Titanium dioxide — A review", *J. Coat. Technol*, (1997), **69**, 59–72.

[37] Brennan JD, Benjamin D, DiBattista E and Gulcev MD, "Using sugar and amino acid additives to stabilize enzymes within sol-gel derived silica", *Chem. Mater.*, (2003), **15**, 737–745.

[38] Brinker CJ and Scherer GW, *Sol-Gel Science — The Physics and Chemistry of Sol-Gel Processing*, Academic Press, Inc., CA, (1990).

[39] Bronshtein A, Aharonson N, Avnir D, Turniansky A and Altstein M, "Sol-gel matrixes doped with atrazine antibodies: Atrazine binding properties", *Chem. Mater.* (1997), **9**, 2632–2639.

[40] Brook MA, Chen Y, Guo K, Zhang Z and Brennan JD, "Sugar modified silanes: precursors for silica monoliths", *J. Mater. Chem*, (2004), **14**, 1469–1479.

[41] Brown W and Foote CS, *Organic Chemistry 3rd Ed.*, Thomas Learning Inc., (2002).

[42] Brunick J, DiNatale C, Bungaro F, Davide F, D'Amico A, Paolesse R, Boschi T, Faccio M and Ferri G, "The application of metalloporphyrins as coating material for quartz microbalance-based chemical sensors", *Anal. Chim. Acta*, (1996), **325**, 53.

[43] Brunswick Laboratories, http://brunswicklabs.com/status_assay.shtml, 2009.

[44] Buchko CJ, *et al.*, "Processing and microstructural characterization of porous biocompatible protein polymer thin films", *Polymer*, (1999), **40**, 7397–7407.

[45] Buer, A, Ugbolue SC and Warner SB, "Electrospinning and properties of some nanofibers", *Textile Research Journal*, (2001), **71**, 323–328.

[46] Buisson P, Hernandez C, Pierre M and Pierre AC, "Encapsulation of lipase in aerogels", *Journal of Non-Crystalline Solids*, (2001), **285**, 295–302.

[47] Byung-Soo Kim and David JM, "Development of biocompatible synthetic extracellular matrices for tissue engineering", *Trends in Biotechnology*, (2001), **16**, 224–230.

[48] Carotta MC, Ferroni M, Gnani D, Guidi V, Merli M, Martinelli G, Casale MC and Notaro M, "Nanostructured pure and Nb-doped TiO_2 as thick film gas sensors for environmental monitoring", *Sens. Actuators B, Chem.*, (1999), **58**, 310–317.

[49] Carturan G, Dal Toso R, Boninsegna S and Dal Monte R , "Encapsulation of functional cells by sol-gel silica: Actual progress and perspective for cell therapy", *J. Mater. Chem.*, (2004), **14**, 2087–2098.

[50] Caruso, RA, Schattka JH and Greiner A,"Titanium dioxide tubes from sol-gel coating of electrospun polymer fibers", *Advanced Materials*, (2001), **13**, p. 1577.

[51] Caturan G, Dal Toso R, Boninsegna S and Dal Monte R, "Encapsulation of functional cells by silica: Actual progress and perspective for cell therapy", *J. Mater. Chem.*, (2004), **14**, 2087–2098.

[52] Cha JN, Shimizu K, Zhou Y, Christiansen SC, Chmelka BF, Stucky GD and Morse DE, "Silicatein filaments and subunits from a marine sponge direc the polymerization of silica and silicones *in vitro*", *Proc. Natl. Acad. Sci. USA*, (1999), **96**, 361–365.

[53] Chen D, Cao Y, Liu B and Kong J, "A BOD biosensor based on a microorganism immobilized on an al2o3 sol-gel matrix", *Analytical and Bioanalytical Chemistry*, (2002), **372**, 737–739.

[54] Chen J, Xu Y, Xin J, Li S, Xia C and Cui J, "Efficient immobilization of whole cells of methylomonas Sp. Strain GYJ3 by sol–gel entrapment", *Journal of Molecular Catalysis B: Enzymatic*, (2004), **30**, 167–172,

[55] Chen X, Cheng G and Dong S, "Amperometric tyrosinase biosensor based on a sol-gel-derived titanium oxide–copolymer composite matrix for detection of phenolic compounds", *Analyst*, (2001), **126**, 1728–1732.

[56] Chen Y, Zhang Z, Sui X, Brennan JD and Brook MA, "Reduced shrinkage of sol-gel derived silicas using sugar-based silsesquioxane precursors", *J. Mater. Chem.*, (2005), **15**, 3132–3141.

[57] Chen YJ, Nie L, Xue XY, Wang YG and Wang TH, "Linear ethanol sensing of SnO_2 nanorods with extremely high sensitivity", *Appl. Phys. Lett.*, (2006), **88**, 083105–1 – 083105–3.

[58] Chen YJ, Xue XY, Wang YG and Wang TH, "Synthesis and ethanol sensing characteristics of single crystalline SnO_2 nanorods", *Appl. Phys. Lett.*, (2005), **87**, 233503–233505.

[59] Cheung MS, Thirumalai D, "Nanopore-protein interactions dramatically alter stability and yield of the native state in restricted spaces", *J. Mol. Biol.*, (2006) **357**, 632–643.

[60] Chia S, Urano J, Tamanoi F, Dunn B and Zink JI, "Patterned hexagonal arrays of living cells in sol-gel silica films", *J. Am. Chem. Soc.*, (2000), **122**, 6488–6489.

[61] Cho Y and Han S, "Catalytic activities of glass-encapsulated horseradish peroxidase at extreme pHs and temperatures", *Bull. Korean Chem. Soc.*, (1999), **20**, 1363–1364.

[62] Choi HN, Kim M and Lee W, "Amperometric glucose biosensor based on sol-gel-derived metal oxide/nafion composite films", *Analytica Chimica Acta*, (2005), **537**, 179–187.

[63] Chopra KL, Major S and Pandya DK, "Transparent conductors — a status review", *Thin Solid Films*, (1983), **102**, 1–46.

[64] Chu XF, Jiang DL, Djurisic AB and Yu HL, "Gas-sensing properties of thick film based on ZnO nano-tetrapods", *Chem. Phys. Lett.*, (2005), **401**, 426–429.

[65] Chu XF, Wang CH, Jiang DL and Zheng CM, "Ethanol sensor based on indium oxide nanowires prepared by carbothermal reduction reaction", *Chem. Phys. Lett.*, (2004), **399**, 461–464.

[66] Coeffier A, Coradin T, Roux C, Bouvet OMM and Livage J, "Sol-gel encapsulation of bacteria: A comparison between alkoxide and aqueous route", *J. Mater. Chem.*, (2001), **11**, 2039–2044.

[67] Colvert JR, Kropp BP, Cheng EY, Pope JC, Brock JW, Adams MC, Austin P, Furness PD and Koyle MA, "The use of small intestinal submucosa as an off-the-shelf urethral

sling material for pediatric urinary incontinence", *J. of Urology*, (2002), **168** (Supplement), 1872–1876.

[68] Comini E, Faglia G, Sberveglieri G, Cantalini C, Passacantando M, Santucci S, Li Y, Wlodarski W and Qu W, "Carbon monoxide response of molybdenum oxide thin films deposited by different techniques", *Sensors and Actuators B*, (2000), **68**, 167–174.

[69] Comini E, Faglia G, Sberveglieri G, Pan Z and Wang ZL, "Stable and highly sensitive gas sensors based on semiconducting oxide nanobelts", *App. Phys. Lett.*, (2002), **81**, 1869–1871.

[70] Comini E, Yubao L, Brando Y and Sberveglieri G, "Gas sensing properties of MoO_3 nanorods to CO and CH_3OH", *Chem. Phys. Lett.*, (2005), **407**, 368–371.

[71] Comini E, Guidi V, Malagu C, Martnelli G, Pan Z, Sberveglieri G and Wang ZL,"Electrical properties of tin oxide twodimensional nanostructures", *J. Phys. Chem. B*, (2004), **108**, 1882.

[72] Comini E, Faglia G, Sberveglieri G, Calestani D, Zanotti L and Zha M, "Tin oxide nanobelts electrical and sensing properties", *Sens. Actuators B*, (2005), 111–112, 2–6.

[73] Comini E, Faglia G, Sberveglieri G, Pan Z and Wang ZL, "Stable and high-sensitive gas sensors based on semiconducting oxide nanobelts", *Appl. Phys. Lett.*, (2002), **81**, 1869–1871.

[74] Coronado JM, Kataoka S, Tejedor-Tejedor I and Anderson MA,"Dynamic phenomena during the photocatalytic oxidation of ethanol and acetone over nanocrystalline TiO_2: Simultaneous FTIR analysis of gas and surface species", *J. Catal.*, (2003), **219**, 219–230.

[75] Cross RBM, De Souza MM and Sankara Narayanan EM, "A low temperature combination method for the production of ZnO nanowires", *Nanotechnology*, (2005), **16**, 2188–2192.

[76] Dai HQ *et al.*, "A novel method for preparing ultra-fine alumina-borate oxide fibres via an electrospinning technique", *Nanotechnology*, (2002), **13**, 674–677.

[77] Dai ZR, Gole JL, Stout JD and Wang ZL, "Tin oxide nanowires, nanoribbons, and nanotubes", *J. Phys. Chem. B*, (2002), **106**, 1274–1279.

[78] Das TK, Khan I, Rousseau DL and Friedman JM, "Temperature dependent quaternary state relaxation in sol-gel encapsulated hemoglobin", *Biospectroscopy*, (1999), **5**, S64–S70.

[79] Dave BC, Soyez H, Miller JM, Dunn B, Valentine JS and Zink ZI, "Synthesis of protein doped sol-gel SiO_2 thin films: Evidence for rotational mobility of encapsulated cytochrome c", *Chem. Mater.*, (1995), **7**, 1431–1434.

[80] Deitzel JM, *et al.*, "The effect of processing variables on the morphology of electrospun nanofibers and textiles", *Polymer*, (2001), **42**(1), 261–272.

[81] Demir MM, *et al.*, "Electrospinning of polyurethane fibers", *Polymer*, (2002), **43**, 3303–3309.

[82] Dersch R, *et al.*, "Electrospun nanofibers: Internal structure and intrinsic orientation", *Journal of Polymer Science Part a-Polymer Chemistry*, (2003), **41**, 545–553.

[83] Dharmaraj N, *et al.*, "Nickel titanate nanofibers by electrospinning", *Materials Chemistry and Physics*, (2004), **87**, 5–9.

[84] Dharmaraj N, *et al.*, "Preparation and morphology of magnesium titanate nanofibres via electrospinning", *Inorganic Chemistry Communications*, (2004), **7**, 431–433.

[85] DiNatale C, Macagnano A, Martinelli E, Paolesse R, D'Arcangelo G, Roscioni C, Finazzi-Agro A and D'Amico A, "Lung cancer identification by the analysis of breath by means of an array of non-selective gas sensors", *Biosens. Bioelectron.*, (2003), **18**, p. 1209.

[86] Ding B, "Fabrication of blend biodegradable nanofibrous nonwoven mats via multi-jet electrospinning", *Polymer*, (2004), **45**, 1895–1902.

[87] Doody MA, Baker GA, Pandey S and Bright FV, "Affinity and mobility of polyclonal anti-dansyl antibodies sequestered within sol-gel-derived biogels", *Chem. Mater.*, (2000), **12**, 1142–1147.

[88] Doong R and Tsai H, "Immobilization and characterization of sol-gel-encapsulated acetylcholinesterase fiber-optic biosensor", *Analytica Chimica Acta*, (2001), **434**, 239–246.

[89] Doshi J and Darrell H. Reneker, "Electrospinning process and application of electrospun fibers", *Journal of Electrostatics*, (1995), **35**, 151–160.

[90] Douglas T and Young M, "Virus particles as templates for materials synthesis", *Adv. Mater.*, (1999), **11**, 679–681.

[91] Drew C, *et al.*, "Metal oxide-coated polymer nanofibers", *Nano Letters*, (2003), **3**, 143–147.

[92] Drexler E, *Engines of Creation — The Coming of Era of Nanotechnology*, Anchor Books (1986).

[93] Droghetti E and Smulevich G, "Effect of sol-gel encapsulation on the unfolding of ferric horse heart cytochrome C", *J. Bio Inorg. Chem.*, (2005), **10**, 696–703.

[94] Dunn B and Zink JI, "Probes of pore environment and molecule-matrix interactions in sol-gel materials", *Chem. Mater.*, (1997), **9**, 2280–2291.

[95] Dutta PK, Ginwalla A, Hogg B, Patton BR, Chiewroth B, Liang Z, Gouma P, Mills M and Akbar S, "Interaction of CO with anatase surfaces at high temperatures: Optimization of a co sensor", *J. Phys. Chem. B*, (1999), **103**, 4412.

[96] Dzenis Y, "Spinning continuous fibers for nanotechnology", *Science*, (2004), **304**, 1917–1919.

[97] Edelmann A and Lendl B, "Towards the optical tongue: Flow-through sensing of tannin-protein interactions based on ftir-spectroscopy", *J. Am. Chem. Soc.*, (2002), **124**, 14741.

[98] Edmiston PL, Wambolt CL, Smith MK and Saavedra SS, "Spectroscopic characterization of albumin and myoglobin entrapped in bulk sol-gel glasses", *Journal of Colloid and Interface Sciences*, (1994), **163**, 395–406.

[99] Eggers DK and Valentine JS, "Crowding and hydration effects on protein conformation: A study with sol-gel encapsulated proteins", *J. Mol. Biol.*, (2001), **314**, 911–922.

[100] Eggers DK and Valentine JS, "Molecular confinement influences protein structure and enhances thermal protein stability", *Protein Sci.*, (2001), **10**, 250–261.

[101] El-Amin SF, Botchwey E, Tuli R, Kofron MD, Mesfin A, Sethuraman S, Tuan RS and Laurencin CT, "Human osteoblast cells: Isolation, characterization, and growth on polymers for musculoskeletal tissue engineering", *J. Biomed. Mater. Res. A*, (2006), **76**, p. 439.

[102] Ellerby LM, Nishida CR, Nishida F, Yamanaka SA, Dunn B, Valentine JS and Zink JI, "Encapsulation of proteins in transparent porous glasses prepared by the sol-gel method", *Science*, (1998), **225**(5048), 1113–1115.

[103] Entcheva E, Bien H, Yin L, Chung C, Farrell M and Kostov Y, "Functional cardiac cell constructs on cellulose-based scaffolding", *Biomaterials*, (2004), **25**, 5753–5762.

[104] Feng P, Wan Q and Wang TH, "Contact-controlled sensing properties of flowerlike ZnO nanostructures", *Appl. Phys. Lett.*, (2005), **87**, 21311–1 – 213111–3.

[105] Fennouh S, Guyon S, Jourdat C, Livage J and Roux C, "Encapsulation of bacteria in silica gels", *C. R. Acad. Sci. Paris, t. 2, Serie II c*, (1999), 625–630.

[106] Ferrer ML, del Monte F and Levy D, "A novel and simple alcohol-free sol-gel route for encapsulation of labile proteins", *Chem. Mater.*, (2002), **14**(9), 3619–3621.

[107] Ferrer ML, Yuste L, Rojo F and del Monte F, "Biocompatible sol-gel route for encapsulation of living bacteria in organically modified silica matrixes", *Chem. Mater.*, (2003), **15**, 3614–3618.

[108] Ferroni M, Guidi V, Martinelli G, Roncarati G, Comini E, Sberveglieri G, Vomiero A and Mea GD, "Coalescence inhibition in nanosized titania films and related effects on chemoresistive properties towards ethanol", *J. Vac. Sci. Technol. B*, (2002), **20**, 523–530.

[109] Ferroni M, Guidi V, Martinelli G, Sacerdoti M, Nelli P and Sberveglieri G, "MoO3 — based sputtered thin films for fast NO2 detection", *Sensors and Actuators B*, (1998), **48**, 285–288.

[110] Ferroni M, Guidi V, Martinelli G, Nelli P, Sacerdoti M and Sberveglieri G, "Characterization of a sputtered molybdenum oxide thin film as a gas sensor", *Thin Solid Films* (1997), **307**, 148–151.

[111] Fertala A, Han WB and Ko FK, "Mapping critical sites in collagen II for rational design of gene-engineered proteins for cell-supporting materials", *Journal of Biomedical Materials Research*, (2001), **57**, 48–58.

[112] Feynman R, A talk given by Dr. Richard Feynman at the annual meeting of American Physical Society, published in the February 1960 issue of Caltech's Engineering Science magazine. Full talk available at: http://www.zyvex.com/nanotech/feynman.html.

[113] Fiandaca G, Vitrano E and Cupane A,"Ferricytochrome c encapsulated in silica nanoparticles: Structural stability and functional properties", *Biopolymers*, (2004), **74**, 55–59.

[114] Finnie KS, Bartlett JR and Woolfrey JL, "Encapsulation of sulfate-reducing bacteria in a silica host", *J. Mater. Chem.*, (2000), **10**, 1099–1101.

[115] Fischer T and Hampp NA, "Encapsulation of purple membrane patches into polymeric nanofibers by electrospinning", *IEEE Transactions on Nanobioscience*, (2004), **3**, 118–120.

[116] Fisher-Wilson J, *The Scientist*, Dec. 10, (2001), p. 22.

[117] Flora K and Brennan JD, "Fluorometric detection of Ca2+ based on an induced change in the conformation of sol-gel entrapped parvalbumin", *Anal. Chem.*, (1998), **70**, 4505–4513.

[118] Flora KK, Dabrowski MA, Musson SP and Brennan JD, C "The effect of preparation and aging conditions on the internal environment of sol-gel derived materials

as probed by 7-azaindole and pyranine fluorescence", *Canadian Journal of Chemistry*, (1999), **77**, 11617–11625.

[119] Folkman J and Moscona A. "Role of cell shape in growth control", *Nature*, (1978), **273**, p. 345.

[120] Fong H, Chun I and Reneker DH, "Beaded nanofibers formed during electrospinning", *Polymer*, (1999), **40**, 4585–4592.

[121] Fong H and Reneker DH, *Elastomeric Nanofibers of Styrene-Butadiene-Styrene Triblock Copolymer*, John Wiley & Sons Inc., (1999).

[122] Fong H, Chun I and Reneker DH. *Beaded Nanofibers Formed During Electrospinning*, Elsevier Sci. Ltd., (1999).

[123] Formhals A, "Process and apparatus for preparing artificial threads", U.S. Patent No. 1975504, (1934).

[124] Francioso L, Forleo A, Capone S, Epifani M, Taurino AM and Siciliano P, "Nanostructured In_2O_3 SnO_2 sol-gel thin film as material for NO_2 detection", *Sens. Actuators B* (2006), **114**, 646–655.

[125] Frenkel-Mullerad H and Avnir D, "Sol-gel materials as efficient enzyme Protectors: Preserving the activity of phosphatases under extreme pH conditions", *J. Am. Chem. Soc.*, (2005), **127**(22), 8077–8081.

[126] Frenot A and Chronakis IS, "Polymer nanofibers assembled by electrospinning", *Current Opinion in Colloid & Interface Science*, (2003), **8**, 64–75.

[127] Friedrikh SV, "Controlling the fiber diameter during electrospinning", *Physical Review Letters*, (2003), **90**, p. 4.

[128] Friedel M, "Effects of confinement and crowding on the thermodynamics and kinetics of a minimalist β-barrel protein", *J. Chem. Phys.*, (2003), **118**, 8106–8113.

[129] Fujihara K, Kotaki M and Ramakrishna S, "Guided bone regeneration membrane made of polycaprolactone/calcium carbonate composite nano-fibers", *Biomaterials*, (2005), **26**, 4139–4147.

[130] Gadre S, Ramachandran K and Gouma PI, "Chemomechanical actuation using PANI hybrids", unpublished work, SUNY Stony Brook, (2006).

[131] Galatsis K, Li XY, Wlodarski W, Comini E, Faglia G and Sberveglieri G, "Semiconductor MoO_3-TiO_2 thin film gas sensors", *Sensors Actuators B*, (2001), **77**, 472–477.

[132] Galatsis K, Li XY, Wlodarski W, Comini E, Sberveglieri G, Cantalini C, Santucci S and Passacantando M, "Comparison of single and binary oxide MoO_3, TiO_2 and WO_3 sol-gel gas sensors", *Sensors and Actuators B*, (2002), **83**, 276–280.

[133] Galatsis K, Li XY, Wlodarski W and Kalantar-Zadeh K, "Sol-gel prepared MoO_3-WO_3 thin films for O_2 gas sensing", *Sensors and Actuators B*, (2001), **77**, 478–483.

[134] Gao JB, Sansinema JM and Wang HL, "Chemical vapor driven polyniline sensor/actuators", *Synthetic Metal*, (2003), **138**, 391–398.

[135] Gao T and Wang T, "Sonochemical synthesis of SnO_2 nanobelt/CdS nanoparticle core/shell heterostructures", *Chem. Comm.*, (2004), **22**, 2558–2559.

[136] Gao T and Wang TH, "Synthesis and properties of multipod-shaped ZnO nanorods for gas-sensor applications", *Appl. Phys. A*, (2005), **80**, 1451–1454.

[137] Gao PX, Lao CS, Hughes WL and Wang ZL, "Three-dimensional interconnected nanowire networks of ZnO", *Chem. Phys. Lett.*, (2005), **408**, 174–178.

[138] Gao T, Li Q and Wang T, "Sonochemical synthesis, optical properties, and electrical properties of core/shell-type ZnO nanorod/CdS nanoparticle composites", *Chem. Mater.*, (2005), **17**, 887–892.

[139] García Sánchez F, Navas Díaz A, Ramos Peinado MC and Belledone C, "Free and sol-gel immobilized alkaline phosphatase-based biosensor for the determination of pesticides and inorganic compounds", *Analytica Chimica Acta*, (2003), **484**, 45–51.

[140] Gardner JW and Bartlett PN, *Electronic Noses: Principles and Applications*, Oxford University Press, NY, (1999).

[141] Garzella C, Bontempi E, Depero LE, Vomiero A, Della Mea G and Sberveglieri G, "Novel selective ethanol sensors: W/TiO_2 sensors by sol-gel spin coating", *Sensors and Actuators B*, (2003), **93**, 495–502.

[142] Gaudette GR, Todato J and Krukenkamp IB, *et al.*, "Computer aided speckle interferometry: A technique for measuring deformation of the surface of the heart", *Annals Biomed. Eng.*, (2001), **29**, 775–780.

[143] Geoffrey MC, *The Cell: A Molecular Approach*, (*Second Edition*), ASM Press, Washington, DC. Sinauer Associates, Inc., Sunderland, Massachusetts, (2000).

[144] Gill I, "Bio-doped nanocomposite polymers: Sol-gel bioencapsulates", *Chem. Mater.*, (2001), **13**, 3404–3421.

[145] Glezer V and Lev O, "Sol-gel vanadium pentoxide glucose biosensor", *J. Am. Chem. Soc.*, (1993), **115**, 2533–2534.

[146] Goring GLG and Brennan JD, "Fluorescence and physical characterization of sol-gel-derived nanocomposite films suitable for the entrapment of biomolecules", *J. Mater. Chem.*, (2002), **12**, 3400–3406.

[147] Gottfried DS, Kagan A, Hoffman BM and Friedmann JM, "Impeded rotation of a protein in a sol-gel matrix", *J. Phys. Chem. B*, (1999), **103**, 2803–2807.

[148] Gouma PI, Bishop A and Iyer KK, "Single crystal metal oxide nanowires as bio-chem sensing probes", *Proceedings of the 6th East Asia Conference on Chemical Sensing*, (2005).

[149] Gouma PI, Mills MJ and Sandhage KH, "Fabrication of free-standing titania-based gas sensors by the oxidation of metallic titanium foils", *J. Am. Ceram. Soc.*, (2000), **83**, 1007–1009.

[150] Gouma PI, "Nanostructured oxide-based selective gas sensor arrays for chemical monitoring and medical diagnostics in isolated environments", *Habitation Journal*, (2005), **10**(2) 99–104.

[151] Gouma PI, Comini E, and Sberveglieri G, "Sol-gel processed MoO_3 and WO_3 thin films for use as selective chemosensors", in *Proc. SPIE Int. Symp. on Microelectronics, MEMS, and Nanotechnology, Vol. 5275*, Eds: Nicolau DV and Miller UR, Dell JM, (2004), 68–75.

[152] Gouma PI, Dutta PK and Mills MJ, "Structural stability of titania thin films", *Nanostruct. Mater.*, (1999), **11**, 1231.

[153] Gouma PI, "Nanostructured polymorphic oxides for advanced chemosensors", *Rev. Adv. Mater. Sci.*, (2003), **5**, 123.

[154] Gouma PI, Iyer KK and Jha PK, "Novel bio-composites for biosensors based on resistive changes", *Proceedings of IEEE Sensors*, (2005).

[155] Gouma PI, Prasad AK and Iyer KK, "Selective nanoprobes for signaling gases", *Nanotechnology*, **17**, pp. S48–S53, (2006).

[156] Gouma PI, Bishop A and Iyer KK, *Proceedings of the 6th East Asia Conference on Chemical Sensing*, (2005).

[157] Gouma PI, "Nanostructured polymorphic oxides for advanced chemosensors", *Rev. Adv. Mater. Sci.*, (2003), **5**, 123–138.

[158] Granqvist CG, *Handbook of Inorganic Electrochromic Materials*, Elsevier, Amsterdam, (1995).

[159] Grantmakher VF, Golubkov MV, Dolgopolov VT, Tsydynzhapov GE and Shashkin AA, "Destruction of localized electron pairs above the magnetic-field-driven superconductor-insulator transition in amorphous In-O films", *JETP Lett.*, (1998), **68**, 363–369.

[160] Gregory O J, Luo Q, Bienkiewicz JM, Erwin BM and Crisman EE, "An apparent n to p transition in reactively sputtered indium–tin-oxide high temperature strain gages", *Thin Solid Films*, (2002), **405**, 263–269.

[161] Gu F, Wang SF, Song CF, Lü MK, Qi YX, Zhou GJ, Xu D and Yuan DR, "Synthesis and luminescence properties of SnO_2 nanoparticles", *Chem. Phys. Lett.*, (2003), **372**, 451–454.

[162] Guan HY, *et al.*, "A novel method for making CuO superfine fibres via an electrospinning technique", *Inorganic Chemistry Communications*, (2003), **6**, 1409–1411.

[163] Guan HY, "A novel method for preparing Co_3O_4 nanofibers by using electrospun PVA/cobalt acetate composite fibers as precursor", *Materials Chemistry and Physics*, (2003), **82**, 1002–1006.

[164] Guan HY, *et al.*, "Fabrication of $NiCO_2O_4$ nanofibers by electrospinning", *Solid State Communications*, (2004), **131**, 107–109.

[165] Guan HY, *et al.*, "Preparation and characterization of NiO nanofibres via an electrospinning technique", *Inorganic Chemistry Communications*, (2003), **6**, 1302–1303.

[166] Gupta P and Wilkes GL, "Some investigations on the fiber formation by utilizing a side-by-side bicomponent electrospinning approach", *Polymer*, (2003), **44**, 6353–6359.

[167] Gurlo A, Ivanovskaya M, Barsan N, Schweizer-Berberich M, Weimar U, Göpel W and Diéguez A, "Grain size control in nanocrystalline In_2O_3 semiconductor gas sensors", *Sens. Actuators B*, (1997), **44**, 327–333.

[168] Gurlo A, Ivanovskaya M, Pfau A, Weimar U and Göpel W, "Sol-gel prepared In_2O_3 thin films", *Thin Solid Films*, (1997), **307**, 288–293.

[169] Hamaguchi T, Yabuki N, Uno M, Yamanaka S, Egashira M, Shimizu Y and Hyodo T, "Synthesis and H_2 gas sensing properties of tin oxide nanohole arrays with various electrodes", *Sens. Actuators B*, (2006), **113**, 852–856.

[170] Han D, Goldgraben S, Frame MD and Gouma PI, "A novel nanofiber scaffold by electrospinning and its utility in microvascular tissue engineering", *Proc. Mat. Res. Soc. Symp.*, (2005), **845**.

[171] Harris LA, "A titanium-dioxide hydrogen detector", *J. Electrochem. Soc.*, (1980), **127**, 2657–2652.

[172] Heinrich VE, "The surfaces of metal oxides", *Rep. Prog. Phy.*, (1985), **48**, 1481–1541.

[173] Henderson MA, "Acetone chemistry on oxidized and reduced $TiO_2(1\ 1\ 0)$", *J. Phys. Chem. B*, (2004), **108**, 18932–18941.

[174] Henderson MA, "Photooxidation of acetone on $TiO_2(1\ 1\ 0)$: Conversion to acetate via methyl radical ejection", *J. Phys. Chem. B*, (2005), **109**, 12062–12070.

[175] Hendricks CD, "Photomicrography of electrically sprayed heavy particles", *AIAA Journal*, (1964), **2**, 733–737.

[176] Hohman MM, "Electrospinning and electrically forced jets. I. Stability theory", *Physics of Fluids*, (2001), **13**, 2201–2220.

[177] Hohman MM, *et al.*, "Electrospinning and electrically forced jets. II. Applications", *Physics of Fluids*, (2001), **13**, 2221–2236.

[178] Hsu CM and Shivkumar S, "Nano-sized beads and porous fiber constructs of poly(epsilon-caprolactone) produced by electrospinning", *Journal of Materials Science*, (2004), **39**, 3003–3013.

[179] Huang Y, Shan W, Liu B, Liu Y, Zhang Y, Zhao Y, Lu H, Tang Y and Yang P, "Zeolite nanoparticle modified microchip reactor for efficient protein digestion", *Lab Chip*, (2006), **6**, 534–539.

[180] Huang J, Matsunaga N, Shimanoe K, Yamazoe N and Kunitake T, "Nanotubular SnO_2 templated by cellulose fibers: Synthesis and gas sensing", *Chem. Mater.*, (2005), **17**, 3513–3518.

[181] Huang L, *et al.*, "Generation of synthetic elastin-mimetic small diameter fibers and fiber networks", *Macromolecules*, (2000), **33**, 2989–2997.

[182] Huang MT, Wu Y, Feick H, Tran N, Weber E and Yang P, "Catalytic growth of sinc oxide nanowires by vapor transport", *Adv. Mater.*, (2001), **13**, 113–116.

[183] Hunter A, Archer CW, Walker PS and Blunn GW, "Attachment and proliferation of osteoblasts and fibroblasts on biomaterials for orthopedic use", *Biomaterials*, (1995), **16**, 287.

[184] Idriss H and Seebauer EG, "Reactions of ethanol over metal oxides", *J. Mol. Catal. A: Chem.*, (2000), **152**, 201–212.

[185] Iijima S, "Helical microtubules of graphitic carbon", *Nature*, (1991), **354**, 56–58.

[186] Iler RK, *The Chemistry of Silica*, John Wiley & Sons, New York, (1979).

[187] Imawan C, Steffes H, Solzbacher F, Obermeier E, "A new preparation method for sputtered MoO_3 multilayers for application in gas sensors", *Sensors and Actuators B*, (2001), **78**, 119–125,

[188] Imawan C, Steffes H, Solzbacher F and Obermeier E, "Structural and gas-sensing properties of V_2O_5-MoO_3 thin films for H_2 detection", *Sensors and Actuators B*, (2001), **77**, 346–351.

[189] Inama L, Dire S, Caturan G and Cavazza A, "Entrapment of viable microorganisms by SiO_2 sol-gel layers on glass surfaces: trapping, catalytic performance and immobilization durability of saccharomyces cerevisiae", *Journal of Biotechnology*, (1993), **30**, 197–210.

[190] Ishibashi S, Higuchi Y, Oa Y and Nakamura K, "Low resistivity indium tin oxide transparent conductive films. I", *J. Vac. Sci. Technol.*, *A*, (1990), **8**, 1399–1402.

[191] Ishizuka N, Kobayashi H, Minakuchi H, Nakanishi K, Hiraoa K, Hosoyab K, Ikegamib T and Tanakab N, "Monolithic silica columns for high-efficiency separations by high-performance liquid chromatography", *Journal of Chromatography A*, (2002), **960**, 85–96.

[192] Jarzebski ZM, " Preparation and physical-properties of transparent conducting oxide-films", *Phys. Status Solidi A*, (1982), **73**, 13–41.

[193] Jason PH, Rae DR, Robert ST and Stephen FB, "Retention of endothelian cell adherence to porcine-derived extracellular matrix after disinfection and sterilization", *Tissue Engineering*, (2002), **8**, 225–234.

[194] Jean DS, "Tissue engineering and reparative medicine", *Annals of the New York Academy of Sciences*, (2002), **961**, 1–9.

[195] Jeong MC, Oh BY, Nam OH, Kim T and Myoung JM, "Three-dimensional ZnO hybrid nanostructures for oxygen sensing application", *Nanotechnology*, (2006), **17**, 526–530.

[196] Jiang HL, *et al.*, "Preparation and characterization of ibuprofen-loaded poly(lactide-co-glycolide)/poly(ethylene glycol)-g-chitosan electrospun membranes" *Journal of Biomaterials Science-Polymer Edition*, (2004), **15**, 279–296.

[197] Jin HJ, *et al.*, "Human bone marrow stromal cell responses on electrospun silk fibroin mats", *Biomaterials*, (2004), **25**, 1039–1047.

[198] Jordan JD, Dunbar RA and Bright FV, "Dynamics of acryodan-labeled bovine and human serum albumin entrapped in a sol-gel derived biogel", *Anal. Chem.*, (1995), **67**, 2436–2433.

[199] Kaciulis S, Pandolfi L, Viticoli S, Sberveglieri G, Zampiceni E, Wlodarski W and Galatsis KLi YX, "Investigation of thin films of mixed oxides for gas sensing applications", *Surface and Interface Analysis*, (2002), **34**, 672–676.

[200] Kadkinova NE and Kostic NM, "Sol-gel glass is not necessarily an inert matrix for enzyme encapsulation catalysis of sulfoxidation by silica", *Journal of Non Crystalline Solids*, (2001), **283**, 63–68.

[201] Kadnikova EN. and Kostic NM, "Oxidation of ABTS by hydrogen peroxide catalyzed by horseradish peroxidase encapsulated into sol-gel glass. Effects of glass matrix on reactivity", *Journal of Molecular Catalysis B: Enzymatic*, (2002), **18**(1), 39–48.

[202] Kaneto K, Kaneko M, Min Y and MacDiarmid AG, "Artificial muscle: Electromechanical actuators using polyaniline films", *Synthetic Metals*, (1995), **71**, 2211–2212.

[203] Kang BS, Heo YS, Tien LC, Norton DP, Ren F, Gila BP and Pearton SJ, "Hydrogen and ozone gas sensing using multiple ZnO nanorods", *Appl. Phys. A*, (2005), **80**, 1029–1032.

[204] Kanungo M and Collison MM, "Controlling diffusion in sol-gel derived monoliths", *Langmuir*, (2005), **21**, 827–829.

[205] Karl T, Prazeller P, Mayr D, Jordan A, Rieder J, Fall R and Lindinger W, "Human breath isoprene and its relation to blood cholesterol levels: New measurements and modeling", *J. Appl. Physiol.*, (2001), **91**, 762–770.

[206] Kato M, Sakai-Kato K, Jin H, Kubota K, Miyano H, Toyoóka T, Dulay MT and Zare RN, "Integration of on-line protein digestion, peptide separation, and protein identification using pepsin-coated photopolymerized sol-gel columns and capillary electrophoresis/mass spectrometry", *Anal. Chem.*, (2004), **76**, 1896–1902.

[207] Kato M, Sakai-Kato K, Matsumoto N and Toyo'oka T, "A protein-encapsulation technique by the sol-gel method for the preparation of monolithic columns for capillary electrochromatography", *Anal. Chem.*, (2002), **74**, 1915–1921.

[208] Kato M, Saruwatari H, Sakai-Kato K and Toyo'oka T, "Silica sol-gel/organic hybrid material for protein encapsulated column of capillary electrochromatography", *Journal of Chromatography A*, (2004), **1044**, 267–270.

[209] Katta P, *et al.*, "Continuous electrospinning of aligned polymer nanofibers onto a wire drum collector", *Nano Letters*, (2004), **4**, 2215–2218.

[210] Kauffmann CG and Mandelbaum RT, "Entrapment of atrazine-degrading enzymes in sol-gel glass", *Journal of Biotechnology*, (1996), **51**, 219–225.

[211] Kawakami K, Sera Y, Sakai S, Ono T and Ijima H, "Development and characterization of a silica monolith immobilized enzyme micro-bioreactor", *Ind. Eng. Chem. Res.*, (2005), **44**, 236–240.

[212] Keeling-Tucker T and Brennan JD, "Fluorescent probes as reporters on the local structure and dynamics in sol-gel-derived nanocomposite materials", *Chem. Mater.*, (2001), **13**, 3331–3350.

[213] Keeling-Tucker T, Rakic M, Spong C and Brennan JD, "Controlling the material properties and biological activity of lipase within sol-gel derived bioglasses via organosilane and polymer doping", *Chem. Mater.*, (2000), **12**, 3695–3704.

[214] Kenawy ER, *et al.*, "Release of tetracycline hydrochloride from electrospun poly(ethylene-co-vinylacetate), poly(lactic acid), and a blend", *Journal of Controlled Release*, (2002), **81**, 57–64.

[215] Kennedy JF, Phillips GO, Wedlock DJ and Williams PA, *Cellulose and its Derivatives: Chemistry, Biochemistry and Applications*, Ellis Horwood Limited, (1985).

[216] Kessick R, Fenn J and Tepper G, "The use of AC potentials in electrospraying and electrospinning processes", *Polymer*, (2004), **45**, 2981–2984.

[217] Khan I, Shannon CF, Dantsker D, Friedman AJ, Perez-Gonzalez-de-Apodaca J and Friedman JM, "Sol-gel trapping of functional intermediates of hemoglobin: Geminate and bimolecular recombination studies", *Biochemistry*, (2000), **39**, 16099–16109.

[218] Kharitonov SA and Barnes PJ, "Exhaled markers of pulmonary disease: State of the art", *American Journal of Respiratory and Critical Care Medicine*, (2001), **163**, 1693–1722.

[219] Kilmov DK, Newfield D and Thirumalai D, "Simulation of β-hairpin folding confined to spherical pores using distributed computing", PNAS, **99**, 8019–8024.

[220] Kim JS and Reneker DH, "Mechanical properties of composites using ultrafine electrospun fibers", *Polymer Composites*, (1999), **20**, 124–131.

[221] Kim YS, Ha SC, Kim K, Yang H, Choi SY, Kim YT, Park JT, Lee CH, Choi J, Paek J and Lee K, "Room-temperature semiconductor gas sensor based on nonstoichiometric tungsten oxide nanorod film", *Appl. Phys. Lett*, (2005), **86**, 213105-1–213107-3.

[222] Klafter J, Drake JM (ed.), *Molecular Dynamics in Restricted Geometrics*, John Wiley, New York, (1989).

[223] Knapp PM, Lingeman JE, Siegel YI, Badylak SF and Demeter RJ, "Biocompatibility of small-intestinal submucosa in urinary tract as augmentation cystoplasty graft and injectable suspension", *J. Endourol.*, (1994), **8**, 125–130.

[224] Kolmakov A and Moskovits M, "Chemical sensing and catalysis by one-dimensional metal oxide nanostructures", *Annu. Rev. Mater. Res.*, (2004), **34**, 151–180.

[225] Kolmakov A, Klenov DO, Lilach Y, Stemmer S and Moskovits M, "Enhanced gas sensing by individual SnO$_2$ nanowires and nanobelts functionalized with Pd catalyst particles", *Nano Lett.*, (2005), **5**, 667–673.

[226] Kolmakov A, Zhang Y, Cheng G and Moskovits M, "Detection of CO and CO$_2$ using tin oxide nanowire sensors", *Adv. Mater.*, (2003), **15**, 997–1000.

[227] Kolmakov A and Moskovits M , "Chemical Sensing and catalysis by one-dimensional metal oxide nanostructures", *Annu. Rev. Mater. Res.*, (2004), **34**, 151–180.

[228] Kong X and Li, Y, "High sensitivity of CuO modified SnO$_2$ Nanoribbons to H$_2$S room temperature", *Sens. Actuators B*, (2005), **105**, 449–453.

[229] Koombhongse S, Liu WX and Reneker DH, "Flat polymer ribbons and other shapes by electrospinning", *Journal of Polymer Science Part B-Polymer Physics*, (2001), **39**, 2598–2606.

[230] Korotcenkov G, Brinzari V, Cerneavschi A, Ivanov M, Golovanov V, Cornet A, Morante J, Cabot A and Arbiol J, "The influence of film structure on In$_2$O$_3$ gas response", *Thin Solid Films*, (2004), **460**, 315–323.

[231] Kress-Rogers E., *Handbook of Biosensors and Electronic Noses: Medicine, Food, and the Environment*, CRC Press, Boca Raton, FL, (1997).

[232] Lan EH, Dave BC, Fukuto JM, Dunn B, Zink JI and Valentine JS, "Synthesis of sol-gel encapsulated heme proteins with chemical sensing properties", *J. Mater. Chem.*, (1999), **9**, 45–53.

[233] Lan EH, Dunn B and Zink JI, "Sol-gel encapsulated anti-trinitrotoluene antibodies in immunoassays for TNT", *Chem. Mater.*, (2000), **12**, 1874–1878.

[234] Lanok J, Santoni TA, Penza M, Loreti S, Menicucci I, Minarinid C and Jelinek M, "Tin oxide thin films prepared by laser-assisted metal–organic CVD: Structural and gas sensing properties", *Surface and Coatings Technology*, (2005), **200**, 1057–1060.

[235] Lang HP, Hegner M, Meyer E and Gerber Ch, "Nanomechanics from atomic resolution to molecular recognition based on atomic force microscopy technology", *Nanotechnology*, (2002), **13**, R29–36.

[236] Larrondo L and Manley RSJ, "Electrostatic fiber spinning from polymer melts. I", *Journal of Polymer Science Part B-Polymer Physics*, (1981), **19**, 909–920.

[237] Lau KT and Hui D, "The revolutionary creation of new advanced materials - carbon nanotube composites", *Composites Part B-Engineering*, (2002), **33**, 263–277.

[238] Lee SW and Belcher AM, "Virus-based fabrication of micro- and nanofibers using electrospinning", *Nano Letters*, (2004), **4**, 387–390.

[239] Li M, Wei Y, MacDiarmid AG and Lelkes PI, "Electrospinning polyaniline-contained gelatin nanofibers for tissue engineering applications", *Biomaterials*, (2006), **27**, 2705–2715.

[240] Li C, Lei B, Zhang D, Liu X, Han S, Tang T, Rouhanizadeh M, Hsiai T and Zhou C, "Chemical gating of In$_2$O$_3$ nanowires by organic and bio molecules", *Appl. Phys. Lett.*, (2003), **83**, 4014–4016.

[241] Li C, Zhang D, Han S, Liu X, Tang T and Zhou, "Diameter-controlled growth of single-crystalline In_2O_3 nanowires and their electronic properties", *Adv. Mater.*, (2003), **15**, 143–146.

[242] Li C, Zhang D, Han S, Liu X, Tang T. Lei B, Liu Z and Zhou C, "Synthesis, electronic properties, and applications of indium oxide nanowires", *Ann. N. Y. Acad. Sci.*, (2003), **1006**, 104–121.

[243] Li C, Zhang D, Lei B, Han S, Liu X and Zhou C, "Surface treatment and doping dependence of in2o3 nanowires as ammonia sensors", *J. Phys. Chem. B.*, (2003), **107**, 12451–12455.

[244] Li C, Zhang D, Liu X, Han S, Tang T, Han J and Zhou C, "In_2O_3 nanowires as chemical sensors", *Appl. Phys. Lett.*, (2003), **82**, 1613–1615.

[245] Li D and Xia YN, "Direct fabrication of composite and ceramic hollow nanofibers by electrospinning", Nano letters, (2004), **4**, 933–938.

[246] Li D and Xia YN, "Fabrication of titania nanofibers by electrospinning", *Nano Letters*, (2003), **3**, 555–560.

[247] Li D, *et al.*, "Photocatalytic deposition of gold nanoparticles on electrospun nanofibers of titania", *Chemical Physics Letters*, (2004), **394**, 387–391.

[248] Li D, Wang YL and Xia YN, "Electrospinning of polymeric and ceramic nanofibers as uniaxially aligned arrays", *Nano Letters*, (2003), **3**, 1167–1171.

[249] Lieber CM, "One dimensional nanostructures-chemistry, physics and applications", *Solid State Communications*, (1998), **107**, 607–616.

[250] Lietti L, Nova I, Ramis G, Acqua LD, Busca G, Giamello E, Forzatti P and Bregani F, "Characterization and reactivity of V2O5-MoO3/TiO2 De-NOx SCR catalysts", *Journal of Catalysis*, (1999), **187**, 419–435.

[251] Lietti L, Alemany JL, Forzatti P, Busca G, Ramis G, Giamello E and Bregani F, "Reactivity of V_2O_5-WO3/TiO2 catalysts in the selective catalytic reduction of nitric oxide by ammonia", *Catal. Today*, (1996), **29**, 143–148.

[252] Likharev KK, "Single-Electron devices and their applications", *Proceedings of the IEEE*, (1999), **87**, 606–632.

[253] Linda G. Griffith, "Reparative medicine: Growing tissues and organs annals of the New York, Emerging design principles in biomaterials and scaffolds for tissue engineering", *Academy of Sciences*, (2002), **961**, 83–95.

[254] Liu B, Cao Y, Chen D, Kong J and Deng J, "Amperometric biosensor based on a nanoporous ZrO_2 matrix", *Analytica Chimica Acta*, (2003), **478**, 59–66.

[255] Liu J, Wang X, Peng Q and Li Y, "Vanadium pentoxide nanobelts: highly selective and stable ethanol sensor materials", *Advanced Materials* (2005), **17**, 764–767.

[256] Liu Y, Zhu W, Tan OK, Yao X and Shen Y, "Structural and gas-sensing properties of nanometre tin oxide prepared by PECVD", *Journal of Materials Science-Materials in Electronics*, (1996), **7**, 279–282.

[257] Liu Z, Deng J and Li D, "A new tyrosinase biosensor based on tailoring the porosity of Al_2O_3 sol-gel to co-immobilize tyrosinase and the mediator", *Analytica Chimica Acta*, (2000), **407**, 87–96.

[258] Liu Z, Liu B, Kong J and Deng J, "Probing trace phenols based on mediator-free alumina sol-gel-derived tyrosinase biosensor", *Anal. Chem.*, (2000), **72**, 4707–4712.

[259] Liu Z, Liu Y, Yang H, Yang Y, Shen G and Yu R, "A mediator-free tyrosinase biosensor based on ZnO sol-gel matrix", *Electroanalysis*, (2005), **17**, 1065–1070.

[260] Liu J, Wang X, Peng Q and Li Y, "Vanadium pentoxide nanobelts: Highly selective and stable ethanol sensor materials", *Adv. Mater.*, (2005), **17**, 764–767.

[261] Liu J, Wang X, Peng Q and Li Y, "Preparation and gas sensing properties of vanadium oxide nanobelts coated with semiconductor oxides", *Sens. Actuators, B*, (2006), **115**, 481–487.

[262] Liu Y, Dong J, Hesketh PJ and Liu MJ, "Synthesis and gas sensing properties of ZnO single crystal flakes", *Mater. Chem.*, (2005), **15**, 2316–2320.

[263] Livage J and Lemerle J, "Transition metal oxide gels and colloids", *Ann. Rev. Mater. Sci.*, (1982), **12**, 103–122.

[264] Livage J. Guzman G, "Aqueous precursors for electrochromic tungsten oxide hydrates", *Solid State Ionics*, (1996), **84**, 205–211.

[265] Loscertales IG, *et al.*, "Electrically forced coaxial nanojets for one-step hollow nanofiber design", *Journal of the American Chemical Society*, (2004), **126**, 5376–5377.

[266] Luckarift HR, Spain JC, Naik RR and O Stone M, "Enzyme immobilization in a biomimetic silica support", *Nature Biotechnology*, (2004), **22**(2), 211–213.

[267] Lutta ST, *et al.*, "Synthesis of vanadium oxide nanofibers and tubes using polylactide fibers as template", *Materials Research Bulletin*, (2005), **40**, 383–393.

[268] Luu YK, *et al.*, "Development of a nanostructured DNA delivery scaffold via electrospinning of PLGA and PLA-PEG block copolymers", *Journal of Controlled Release*, (2003), **89**, 341–353.

[269] M Kato M, Inuzuka K, Sakai-Kato K and Toyo'oka T, "Monolithic bioreactor immobilizing trypsin for high-throughput analysis", *Anal. Chem.*, (2005), **77**, 1813–1818.

[270] Ma D, Li M, Patil AJ and Mann S, "Fabrication of protein/silica core shell nanoparticles by microemulsion based molecular wrapping", *Adv. Mater.*, (2004), **16**(20), 1838–1841.

[271] MacDiarmid AG, *et al.*, "Electrostatically-generated nanofibers of electronic polymers", *Synthetic Metals*, (2001), **119**, 27–30.

[272] Madhugiri S, *et al.*, "Electrospun mesoporous molecular sieve fibers", *Microporous and Mesoporous Materials*, (2003), **63**, 75–84.

[273] Madhugiri S, *et al.*, "Electrospun mesoporous titanium dioxide fibers", *Microporous and Mesoporous Materials*, (2004), **69**, 77–83.

[274] Madler L, Sahm T, Gurlo A, Grunwaldt JD, Barsan N, Weimar U and Pratsinis SE, "Sensing low concentrations of CO using flame-spray-made Pt/SnO_2 nanoparticles", *J. Nanoparticle Res.*, (2006), **8**, 783–796.

[275] Marques AP and Reis RL, "Hydroxyapatite reinforcement of different starch-based polymers affects osteoblast-like cells adhesion/spreading and proliferation", *Mat. Sci. Eng. C-Bio S.*, (2005), **25**, 215.

[276] Martinelli G, Carotta MC, Ferroni M, Sadaoka Y and Traversa E, "Screen-printed perovskite-type thick films for environmental monitoring", *Sensors and Actuators B*, (1999), **55**, 99–110.

[277] Martson M, Jouko V. Timo H. Pekka L. and Pekka S. "Is cellulose sponge degradable or stable as implantation material? An *in vivo* subcutaneous study in the rat", *Biomaterials*, (1999), **20**, 1989–1995.

[278] Mather BA, *The Scientist*, (2002), **38**.

[279] Mazza T, Barborini E, Kholmanov IN, Piseri P, Bongiorno G, Vinati S, Milani P, Ducati C, Cattaneo, Li Bassi DA, Bottani CE, Taurino AM and Siciliano P, "Libraries of cluster-assembled titania films for chemical sensing", *Appl. Phys. Lett.*, (2005), **87**, 103108.

[280] Meijer GJ, de Bruijn JD, Koole R and van Blitterswijk C, "Cell-based bone tissue engineering", *PLoS Med*, (2007), **4**, 260.

[281] Messing GL, Zhang SC and Jayanthi GV, "Ceramic powder synthesis by spray pyrolysis", *J. Am. Ceram. Soc*, (1993), **76**, 2707–2726.

[282] Savage NO, *et al.*, "Titanium dioxide based high temperature carbon monoxide selective sensor", *Sens. Actuators B, Chem.*, (2001), **72**, 239–248.

[283] Montesperelli G, Pumo A, Traversa E, Gusmano G, Bearzotti A, Montenero A and Gnappi G, "Sol-Gel processed TiO2-based thin-films as innovative humidity sensors", *Sens. Actuators B, Chem.*, (1995), **25**, 705–709.

[284] Morrison SR, "Selectivity in semiconductor gas sensors", *Sens. Actuators, B*, (1987), **12**, 425–440.

[285] Moseley PT and Crocker AJ, *Sensor Materials*, Institute of Physics, Bristol, (1996).

[286] Motokawa M, Kobayashi H, Ishizuka N, Minakuchi H, Nakanishic K, Jinnaia H, Hosoyaa K, Ikegamia T and Tanakaa N, "Monolithic silica columns with various skeleton sizes and through-pore sizes for capillary liquid chromatography", *Journal of Chromatography A*, (2002), **961**, 53–63.

[287] Motokawa M, Kobayashi H, Ishizuka N, Minakuchi H, Nakanishic K, Jinnaia H, Hosoyaa K, Ikegamia T and Tanakaa N, "Monolithic silica columns with various skeleton sizes and through-pore sizes for capillary liquid chromatography", *Journal of Chromatography A*, (2002), **961**, 53–63.

[288] Mueller R, Kammler HK, Wegner K and Pratsinis SE, "OH surface density of SiO_2 and TiO_2 by thermogravimetric analysis", *Langmuir*, (2003), **19**, 160–165.

[289] Mutschall D, Holzner K and Obermeier E, "Sputtered Molybdenum oxide thin films for NH_3 detection", *Sens. and Actuators B*, (1996), **35–36**, 320–324.

[290] Mädler L, *et al.*, "Sensing low concentrations of CO using flame-spray-made Pt/SnO_2 nanoparticles", *Journal of Nanoparticle Research*, (2006), **8**, 783–796.

[291] Narang U, Prasad PN and Bright FV, "A novel protocol to entrap active urease in a tetraethoxysilane-derived sol-gel thin-film architecture", *Chem. Mater.*, (1994), **6**, 1596–1598.

[292] Nassif N, Bouvet O, Rager MN, Roux C, Coradin T and Livage J, "Living bacteria in silica gels", *Nature Materials*, (2002), **1**, 42–44.

[293] Nassif N, Coiffier A, Coradin T, Roux C and Livage J, "Viability of bacteria in hybrid aqueous silica gels", *Journal of Sol-Gel Science and Technology*, (2003), **26**, 1141–1144.

[294] Nassif N, Roux C, Coradin T, Bouvet OMM and Livage J, "Bacteria quorum sensing in silica matrices", *J. Mater. Chem.*, (2004), **14**, 2264–2268.

[295] Nassif N, Roux C, Coradin T, Rager MN, Bouvet OMM and Livage J, "A sol-gel matrix to preserve the viability of encapsulated bacteria", *J. Mater. Chem.*, (2003), **13**, 203–208.

[296] Nicoll SB, Radin S, Santost EM, Tuan RS and Ducheyne P, "*In vitro* release kinetics of biologicallyactive transforming growth factor-B1 horn a novel porous glass carrier", *Biomaterials*, (1997), **18**, 853–959.

[297] NOSE II—2nd Network on artificial Olfactory Sensing. Home Page, http://www.nose-network.org (2009).

[298] Carotta MC, Ferroni M, Gnani D, Guidi V, Merli M, Martinelli G, Casale MC and Notaro M, "Nanostructured pure and Nb-doped TiO_2 as thick film gas sensors for environmental monitoring", *Sens. and Actuators B: Chemical*, (1999), **58**, 310–317.

[299] Novak J P, Snow ES, Houser EJ, Park D, Stepnowski JL and Gill RA, "Nerve agent detection using networks of single-walled carbon nanotubes", *Appl. Phys. Lett.*, (2003), **83**, 4026.

[300] Obee TN and Brown RT, "TiO_2 photocatalysis for indoor air applications-effects of humidity and trace contaminant levels on the oxidation rates of formaldehyde, toluene, and 1,3-butadiene", *Environ. Sci. Technol.*, (1995), **29**, 1223–1231.

[301] Okonski CT and Thacher HC, "The distortion of aerosol droplets by an electric field", *Journal of Physical Chemistry*, (1954), **57**, 955–958.

[302] Osmetech Home Page "eNose Technology," http://www.osmetech.co.uk/enose.htm (2009).

[303] Palmer T, *Understanding Enzymes, Fourth Edition*, Prentice Hall/Ellis Horwood, (1995).

[304] Park CO and Akbar SA, "Ceramics for chemical sensing", *Journal of Materials Science*, (2003), **38**, 4611–4637.

[305] Park W I, Yi GC, Kim M and Pennycook SJ, "Quantum confinement observed in ZnO/ZnMgo nanorod heterostructures", *Adv. Mater.*, (2003), **15**, 526–529.

[306] Parthangal PM, Cavicchi RE, Montgomery CB, Turner S and Zachariah MR, "Restructuring tungsten thin films into nanowires and hollow square cross-section microducts", *J. Mater. Res.*, (2005), **20**, 2889–2894.

[307] Phillips M, in Disease Markers in *Exhaled Breath*, N Marczin, SA Kharitonov, MH Yacoub, and PJ Barnes (Eds.), Marcel Dekker, New York, (2002) p. 219.

[308] Pierre AC, "The sol-gel encapsulation of enzymes", *Biocatalysis and Biotrasformation*, (2004), **22**, 145–170.

[309] Ping G, Yuan M, Vallieres M, Dong H, Sun Z and Wei Y, "Effects of confinement on protein folding and protein stability", *J. Chem. Phys.*, (2003), **118**, 8042–8048.

[310] Pioselli B, Bettati S, Demidkina TV, Zakomirdina LN, Phillips RS and Mozzarelli A, "Tyrosine phenol-lyase and tryptophan indole-lyase encapsulated in wet nanoporous silica gels: Selective stabilization of tertiary conformations", *Protein Sci.*, (2004), **13**, 913–924.

[311] Pioselli B, Bettati S and Mozzarelli A, "Confinement and crowding effects on tryptophan synthase $\alpha 2\beta 2$ complex", *FEBS Letters*, (2005), **579**, 2197–2202.

[312] Pokhrel S and Nagaraja KS, "Electrical and humidity sensing properties of molybdenum (VI) oxide and tungsten (VI) oxide composites", *Phys. Stat. Sol (a)*, (2003), **198**, 343–349.

[313] Pokhrel S and Nagaraja KS, "Solid state electrical conductivity and humidity sensing properties of Cr_2O_3-MoO_3 composites", *Phys. Stat. Sol (a)*, (2002), **194**, 140–146.

[314] Polleux J, Gurlo A, Barsan N, Weimar U, Antonietti M and Niederberger M, "Template free synthesis and assembly of single crystalline tungsten oxide nanowires and their gas-sensing properties", *Angew. Chem. Int. Ed.*, (2006), **45**, 261–265

[315] Pope EJA, "Bioartificial organs I: Silica gel encapsulated pancreatic islets for the treatment of diabetes mellitus", *Journal of Sol-Gel Science and Technology*, (1997), **8**, 635–639.

[316] Prasad AK, Kubinski DJ and Gouma PI, "Comparison of sol-gel and ion beam deposited MoO_3 thin films for selective ammonia detection", *Sens. and Actuators B*, (2003), **93**, 25–30.

[317] Prasad AK, Gouma PI, Kubinski DJ, Visser JH, Soltis RE and Schmitz PJ, "Reactively sputtered MoO_3 films for NH_3 sensing", *Thin Solid Films*, (2003), **436**, 46–51.

[318] Prasad AK, Kubinski D and Gouma PI, "Comparison of sol-gel and ion beam deposited MoO_3 thin film gas sensors for selective ammonia detection", *Sens. Actuators, B*, (2003), **93**, 25–30.

[319] Prasad AK , Study of Gas Specificity in MoO_3/WO_3 "Thin film sensors and their arrays", Ph.D. thesis, May (2005), SUNY Stony Brook, NY.

[320] Pratsinis S E, Zhu WH and Vemury S, "The role of gas mixing in flame synthesis of titania powders", *Powder Technol.*, (1996), **86**, 87–93.

[321] Premkumar JR, Rosen R, Belkin S and Lev O, "Sol-gel luminescence biosensors: Encapsulation of recombinant E. coli reporters in thick silicate films", *Analytica Chimia Acta*, (2002), **462**, 11–23.

[322] Pressi G, Dal Toso R and Dal Monte R, "Production of enzymes by plant cells immobilized by sol-gel silica", *J. Sol-Gel Sci. and Tech.*, (2003), **26**, 1189–1193.

[323] Radin S, Ducheyne P, Kamplain T and Tan BH, "Silica sol-gel for the controlled release of antibiotics-i: synthesis, characterization, and *in vitro* release", *Journal of Biomedical Materials Research*, (2001), **57**, 313–320.

[324] Radin S, Falaize S, Lee MH and Ducheyne P, "*In vitro* bioactivity and degradation behavior of silica xerogels intended as controlled release materials", *Biomaterials*, (2002), **15**, 3113–3122.

[325] Raff J, Soltmann U, Matys S, Selenska-Pobell S, Böttcher H and Pompe W, "Biosorption of uranium and copper by biocers", *Chem. Mater.*, (2003), **15**, 240–244.

[326] Raible I, Burghard M, Schlecht U, Yasuda A and Vossmeyer T, "V_2O_5 nanofibres: Novel gas sensors with extremely high sensitivity and selectivity to amines", *Sens. Actuators, B*, (2005), **106**, 730–735.

[327] Ramgir NS, Mulla IS and Vijayamohan KP, "A room temperature nitric oxide sensor actualized from Ru-doped SnO_2 nanowires", *Sens. Actuators B*, (2005), **107**, 708–715.

[328] Rangpukan RRD, Special issue: The Fiber Society Spring 2001 Conference, Raleigh NC, 2001, "Development of electrospinning from molten polymers in vacuum", *J. Textile Apparel, Technol. Manage.*, (2001), **1**

[329] Rao CNR and Cheetham AK, "Science and technology of nanomaterials: Current status and future prospects", *J. Mater. Chem.*, (2001), **11**, 2887–2894.

[330] Rao MS and Dave BC, "Selective intake and release of proteins by organically-modified silica sol-gels", *J. Am. Chem. Soc.*, (1998), **120**, 13270–13271.

[331] Rathore N, Knotts IV TA and de Pablo JJ, "Confinement effects on the thermodynamics of protein folding: Monte Carlo simulations", *Biophysical Journal*, (2006), **90**, 1767–1773.

[332] Ravindra R, Zhao S, Hermann G and Winter R, "Protein encapsulation in mesoporous silicate: The effects of confinement on protein stability, hydration and volumetric properties", *J. Am. Chem. Soc.*, (2004), **126**, 12224–12225.

[333] Ray A, Feng M and Tachikawa H, "Direct electrochemistry and raman spectroscopy of sol-gel encapsulated myoglobin", *Langmuir*, (2005), **21**, 7456–7460.

[334] Reetz MT , Zonta A and Simpelkamp J, "Efficient immobilization of lipases by entrapment in hydrophobic sol-gel materials", *Biotechnology and Bioengineering*, (1996), **49**, 5, 527–534.

[335] Reneker DH, *et al.*, "Bending instability of electrically charged liquid jets of polymer solutions in electrospinning", *Journal of Applied Physics*, (2000), **87**, 4531–4547.

[336] Richard AA and Theodore DB, "Resorbable extracellular matrix grafts in urologic reconstruction", *Int. Braz. J. Urol.*, (2005), **31**, 192–203.

[337] Richard S and Fox CF, Tissue engineering, UCLA Symposia on Molecular and Cellular Biology, New Series, Volume 107, Alan R. Liss, Inc., New York.

[338] Rietti-Shati M, Ronen D and Mandelbaum RT, "Atrazin degradation by pseudomonas strain adp entrapped in sol-gel", *Journal of Sol-Gel Science and Technology*, (1996), **7**, 77–79.

[339] Risby TH and Sehnert SS, "Clinical application of breath biomarkers of oxidative stress status", *Free Radical Biology & Medicine*, (1999), **27**, 1182–1192.

[340] Rout CS, Krishna HS, Vivekchand SRC, Govindraj A and Rao CNR, "Hydrogen and ethanol sensors based on ZnO nanorods, nanowires and nanotubes", *Chem. Phys. Lett.*, (2006), **418**, 586–590.

[341] Roveri N, Morpurgo M, Palazzo B, Parma B and Vivi L, "Silica xerogels as a delivery system for the controlled release of different molecular weight heparins", *Analytical and Bioanalytical Chemistry*, (2005), **381**, 601–606.

[342] Rubenstein DA, "Development of a novel bioassay chamber to optimize autologous endothelial cell viability and density on topological and topographical substrates", Ph.D. Dissertation Stony Brook University.

[343] Rupcich N, Goldstein A and Brennan JD, "Optimization of sol-gel formulations and surface treatments for the development of pin-printed protein microarrays", *Chem. Mater.*, (2003), **15**, 1803–1811.

[344] Sakai-Kato K, Kato M, Nakakuki H and Toyo'oka T, "Investigation of structure and enantioselectivity of bsa encapsulated sol-gel columns prepared for capillary electrochromatography", *Journal of Pharmaceutical and Biomedical Analysis*, (2003), **31**, 299–309.

[345] Sakai-Kato K, Kato M and Toyo'oka T, "On-line trypsin-encapsulated enzyme reactor by the sol-gel method integrated into capillary electrophoresis", *Anal. Chem.*, (2002), **74**, 2943–2949.

[346] Sakai-Kato K, Kato M and Toyo'oka T, "Creation of an on-chip enzyme reactor by encapsulating trypsin in sol-gel on a plastic microchip", *Anal. Chem.*, (2003), **75**, 388–393.

[347] Salgado AJ, Continho OP and Reis RL, "Bone tissue engineering: state of the art and future trends", *Macromol. Biosci.*, (2004), **4**, 743.

[348] Salgado AJ, Gomes ME, Chou A, Coutinho OP, Reis RL and Hutmacher DW, "Preliminary study on the adhesion and proliferation of human osteoblasts on starch-based scaffolds", *Mat Sci Eng C-Bio S*, (2002), **20**, 27.

[349] Samuni U, Navati MS, Juszczak LJ, Dantsker D, Yang M and Friedman JM, "Unfolding and refolding of sol-gel encapsulated carbonmonoxymyoglobin: An orchestrated spectroscopic study of intermediates and kinetics", *J. Phys. Chem. B* (2000), **104**, 10802–10813.

[350] Sanchez C, Julian B, Belleville P and Popall M, "Applications of hybrid organic-inorganic nanocomposites", *J. Mater. Chem.*, (2005), **15**, 3559–3592.

[351] Santos EM, Radin S and Ducheyne P, "Sol-gel derived carrier for the controlled release of proteins", *Biomaterials*, (1999), **20**, 1695–1700.

[352] Savage NO, Akbar SA and Dutta PK, "Titanium dioxide based high temperature carbon monoxide selective sensor", *Sens. and Actuators B: Chemical*, (2001), **72**, 239–248.

[353] Sawicka K and Gouma PI, "Electrospun composite nanofibers for functional applications", *J. Nanoparticle Research*, (2006), **8**, 769-781.

[354] Sawicka KM, Prasad AK and Gouma PI, "Metal oxide nanowires for use in chemical sensing applications", *Sensor Letters*, (2005), **3**, 31–35.

[355] Sawicka KM, Gouma P and Simon S; "Electrospun bio-composite nanofibers for urea biosensing", *Sensors Act. B*, (2005), **108**, 585–588.

[356] Sawicka KM, Prasad AK, Gadia S and Gouma PI, "Processing and characterization of nanostructured metal oxides and nanocomposites for use in bio-chemical sensing applications", *Proc. AIST*, (2004).

[357] Sawicka KM, Prasad AK and Gouma PI, "Metal oxide nanowires for use in chemical sensing applications", *Sensor Letters*, (2005), **3**, 31–35.

[358] Schreuder-Gibson HL, Truong Q, Walker JE, Owens JR, Wander JD and Jones WE, Jr., "Chemical and biological protection and detection in fabrics for protective clothing", *MRS Bull.*, (2003), **28**, 574.

[359] Seiyama T, Kato A, Fujishi K and Nagatani M, "A new detector for gaseous components using semiconductive thin films", *Analytical Chemistry*, (1962), **34**, 1502.

[360] Semancik S and Cavicci RE, "Novel materials and applications of electronic noses and tongues", *Appl. Surf. Sci.*, (1993), **70**, 337.

[361] Shamansky LM, Luong KM, Han D and Chronister EL, "Photoinduced kinetics of bacteriorhodopsin in a dried xerogel glass", *Biosensors and Bioelectronics*, (2002), **17**, 227–231.

[362] Shao CL, *et al.*, "A novel method for making ZrO_2 nanofibres via an electrospinning technique", *Journal of Crystal Growth*, (2004), **267**, 380–384.

[363] Shao CL, *et al.*, "Electrospun nanofibers of NiO/ZnO composite", *Inorganic Chemistry Communications*, (2004), **7**, 625–627.

[364] Shao CL, *et al.*, "Preparation of Mn_2O_3 and Mn_3O_4 nanofibers via an electrospinning technique", *Journal of Solid State Chemistry*, (2004), **177**, 2628–2631.

[365] Shen C and Kostic NM, "Kinetics of photoinduced electron-transfer reactions within sol-gel silica glass doped with zinc cytochrome c. study of electrostatic effects in confined liquids", *J. Am. Chem. Soc.* (1997), **119**, 1304–1312.

[366] Shen G, Chen, P-C, Ryu, K and Zhou C, "Devices and chemical sensing applications of metal oxide nanowires", *Journal of Materials Chemistry*, (2009), **19**, 828–839.

[367] Shevade AV, Ryan MA, Homer ML, Manfreda AM, Zhou H and Manatt KS, "Molecular modeling of polymer composite–analyte interactions in electronic nose sensors", *Sens. Actuators, B, Chem.*, (2003), **93**, 84.

[368] Shieh J, Feng HM, Hong MH and Juang HY, "WO_3 and W-Ti-O thin film gas sensors prepared by sol-gel dip coating", *Sens. and Actuators B*, (2002), **86**, 75–80.

[369] Shimano J and Macdiarmid A, "Polyaniline, a dynamic block copolymer: Key to attaining its intrinsic conductivity?", *Synthetic Metals.*, (2001), **123**, 251.

[370] Shin YM, Hohman MM, Brenner MP and Rutledge GC, "Experimental characterization of electrospinning: The electrically forced jet and instabilities", *Polymer*, (2001), **42**, 9955–9967.

[371] Shin YM, Hohman MM, Brenner MP and Rutledge GC, "Electrospinning: A whipping fluid jet generates submicron polymer fibers", *Appl. Phys. Lett.*, (2001), **78**, 1149.

[372] Shin YM, *et al.*, "Experimental characterization of electrospinning: The electrically forced jet and instabilities", *Polymer*, (2001). **42**, 9955–9967.

[373] Sisk BC and Lewis NS, "Estimation of chemical and physical characteristics of analyte vapors through analysis of the response data of arrays of polymer-carbon black composite vapor detectors", *Sens. Actuators, B, Chem.*, (2003), **96**, 268.

[374] Smith AD, Cowan JO, Filsell S, McLachlan C, Monti-Sheehan G, Jackson P and Taylor DR, "Diagnosing asthma: Comparisons between exhaled nitric oxide measurements and conventional tests", *American Journal of Respiratory and Critical Care Medicine*, (2004), **169**, 473–478.

[375] Smith DM, Scherer GW and Anderson JM, "Shrinkage during drying of silica gel", *Journal of Non-Crystalline Solids*, (1995), **188**, 191–206.

[376] Smith LA, Liu X and Ma PX, "Tissue-engineering with nano-fibrous scaffolds", *Soft Matter*, (2008), **4**, 2144–2149.

[377] Sreenivas K, Kumar S, Choudhary J and Gupta V "Growth of zinc oxide nanostructures", Pramana, (2005), **65**, 809–814.

[378] Statham MJ, Hammett F, Harris B, Cooke RG, Jordan RM and Roche A, "Net-shape manufacture of low-cost ceramic shapes by freeze-gelation", *Journal of Sol-Gel Sci. and Technol.*, (1998), **13**, 171–175.

[379] Stephen FB, "Extracellular matrix as a scaffold for tissue engineering in veterinary medicine: Applications to soft tissue healing", *Clinical Techniques in Equine Practice*, (2004), **3**, 173–181.

[380] Stephen FB, "The extracellular matrix as a scaffold for tissue reconstruction", *Cell & Development Biology*, (2002), **13**, 377–383.

[381] Stetter R, Strathmann S, McEntegart C, DeCastro M and Penrose WR, "New sensor arrays and sampling systems for a modular electronic nose", *Sens. Actuators, B, Chem.*, (2000), **69**, 410.

[382] Strobel R, Stark WJ, Madler L, Pratsinis SE and Baiker A, "Flame-made platinum/alumina: Structural properties and catalytic behaviour in enantioselective hydrogenation", *J. Catal.*, (2003), **213**, 296–304.

[383] Strutt J, "On the equilibrium of liquid conducting masses charged with electricity", *Philos. Mag.*, London, Edinburgh and Dublin, (1882), **14**, 184–186.

[384] Studer SM, Orens JB, Rosas I, Krishman JA, Cope KA, Yang S, Conte JV, Becker PB and Risby TH, "Patterns and significance of exhaled-breath biomarkers in lung transplant recipients with acute allograft rejection", *The Journal of Heart and Lung Transplantation*, (2001), **20**, 1158–1166.

[385] Sui X, Cruz-Aguado JA, Chen Y, Zhang Z, Brook MA and Brennan JD,"Properties of human serum albumin entrapped in sol-gel derived silica bearing covalently tethered sugars", *Chem. Mater.*, (2005), **17**, 1174–1182.

[386] Sui X, Lin T, Tleugabulova D, Chen Y, Brook MA and Brennan JD, "Monitoring the distribution of covalently tethered sugar moieties in sol-gel-based silica monoliths with fluorescence anisotropy: Implications for entrapped enzyme activity", *Chem. Mater.*, (2006), **18**, 887–896.

[387] Suito K, Kawai N and Masuda Y, "High pressure synthesis of orthorhombic SnO_2", *Mater. Res. Bull.*, (1975), **10**, 677–680.

[388] Sun ZC, *et al.*, "Compound core-shell polymer nanofibers by co-electrospinning", *Advanced Materials*, (2003), **15**, 1929.

[389] Sundaray B, *et al.*, "Electrospinning of continuous aligned polymer fibers", *Appl. Phys. Lett.*, (2004), **84**, 1222–1224.

[390] Sunu S S, Prabhu E, Jayaraman V, Gnanasekar KI and Gnanasekaran T, "Gas sensing properties of PLD made MoO_3 films", *Sens. and Actuators B*, (2003), **94**, 189–196.

[391] Surnev S, Ramsey MG and Netzer FP, "Vanadium oxide surface studies", *Prog. Surf. Sci.*, (2003), **73**, 117–165.

[392] Suryanarayana C, "The Structure and properties of nanocrystalline materials — Issues and concerns", *J. of Mater.*, (2002), 24–27.

[393] Sze SM, *Semiconductor Sensors, 1st edition*, John Wiley & Sons Inc., (2004).

[394] Taguchi N, *Japanese Patent Application*, (1962).

[395] Tanakaa N, Kobayashia H, Ishizukab N, Minakuchib H, Nakanishi K, Hosoya K and Ikegami T, "Monolithic silica columns for high-efficiency chromatographic separations", *Journal of Chromatography A*, (2002), **965**, 35–49.

[396] Taurino AM, Forleo A, Francioso L, Siciliano P, Stalder M and Nesper R, "Synthesis, electrical characterization, and gas sensing properties of molybdenum oxide nanorods", *Appl. Phys. Lett.*, (2006), **88**, 152111–1– 152111–3.

[397] Taylor AP, Finnie KS, Bartlett JR and Holden PJ, "Encapsulation of viable aerobic microorganisms *in silica*", *Journal of Sol-Gel Science and Technology*, (2004), **32**, 223–228.

[398] Taylor GI and McEwan AD, "Stability of a horizontal fluid interface in a vertical electric field", *Journal of Fluid Mechanics*, (1965), **22**, 1.

[399] Taylor GI, "Electrically driven jets", *Proc R. Soc. London, Ser. A*, (1969), **313**, 453.

[400] Teleki A, Pratsinis SE, Kalyanasundaram K and Gouma PI, "Sensing of organic vapors by flame-made TiO_2 nanoparticles", *Sensors and Actuators B*, (2006), **119**, 683–690.

[401] Teleki A, Pratsinis SE, Wegner K, Jossen R and Krumeich F, "Flame-coating of titania particles with silica", *J. Mater. Res.*, (2005), **20**, 1336–1347.

[402] Thaler ER, Bruney FC, Kennedy DW and Hanson CW, " Use of an electronic nose to distinuguish cerebrospinal fluid from senim", *Arch. Otorynchology: Head & Neck Surgery*, (2000), **126**, 71–74.

[403] Theron A, Zussman E and Yarin AL, *Electrostatic Field-Assisted Alignment of Electrospun Nanofibres*, IOP Publishing Ltd., (2001).

[404] Theron SA, Zussman E and Yarin AL, "Experimental investigation of the governing parameters in the electrospinning of polymer solutions", *Polymer*, (2004), **45**, 2017–2030.

[405] Tian ZR, Voigt JA, Liu J, McKenzie B, McDermott MJ, Rodriguez MA, Konishi H and Xu H, "Complex and oriented ZnO nanostructures", *Nat. Mater.*, (2003), **2**, 821–826.

[406] Tien LC, Wang HT, Kang BS, Ren F, Sadik PW, Norton DP, Pearton SJ and Lin J, "Room-temperature hydrogen-selective sensing using single Pt-coated ZnO nanowires at microwatt power levels", *Electrochem. Solid-State Lett.*, (2005), **8**, G230–G232.

[407] Tomer V, *et al.*, "Selective emitters for thermophotovoltaics: Erbia-modified electrospun titania nanofibers", *Solar Energy Materials and Solar Cells*, (2005), **85**, 477–488.

[408] Tripathi VS, Kandimalla VB and Ju H, "Preparation of ormosil and its applications in the immobilizing biomolecules", *Sens. and Actuators B*, (2006), **114**, 1071–1082.

[409] Tsai PP, Schreuder-Gibson H and Gibson P, "Different electrostatic methods for making electret filters", *J. Electrostat.*, **54**, (2002), 333–341.

[410] Ulrich Soltmann U, Böttcher H, Koch D and Grathwohl G, "Freeze gelation: A new option for the production of biological ceramic composites (biocers)", *Materials Letters*, (2003), **57**(19), 2861–2865.

[411] Van Lenthe GH, Hagenmuller H, Bohner M, Hollister R, Meinel L and Müller R, "Nondestructive micro-computed tomography for biological imaging and quantification of scaffold–bone interaction *in vivo*", *Biomaterials*, (2007), **28**, 2479.

[412] Vao-Soongnern V and Mattice WL, "Dynamic properties of an amorphous polyethylene nanofiber", *Langmuir*, (2000), **16**, 6757–6758.

[413] Vao-soongnern V, Doruker P and Mattice WL, "Simulation of an amorphous polyethylene nanofiber on a high coordination lattice", *Macromolecular Theory and Simulations*, (2000), **9**, 1–13.

[414] Vemury S and Pratsinis SE, "Dopants in flame synthesis of titania", *J. Am. Ceram. Soc.*, (1995), **78**, 2984–2992.

[415] Vera-Avila LE, Morales-Zamudio E and Garcia-Camacho MP, "Activity and reusability of sol-gel encapsulated alpha-amylase and catalase performance in flow-through systems", *Journal of Sol-Gel Science and Technology*, (2004), **30**, 197–204.

[416] Viswanathamurthi P, *et al.*, "Preparation and morphology of palladium oxide fibers via electrospinning", *Materials Letters*, (2004), **58**, 3368–3372.

[417] Viswanathamurthi P, *et al.*, "Ruthenium doped TiO_2 fibers by electrospinning", *Inorganic Chemistry Communications*, (2004), **7**, 679–682.

[418] Viswanathamurthi P, *et al.*, "Vanadium pentoxide nanofibers by electrospinning", *Scripta Materialia*, (2003), **49**, 577–581.

[419] Wada Y, Suga M, Kure T, Yoshimura T, Sudo Y, Kobayashi T, Goto Y and Kondo S, "Quantum transport in polycrystalline silicon 'slit nano wire'", *Appl. Phys. Lett.*, (1994), **65**, 634–626.

[420] Walser R and van Gunsteren WF, "Viscosity dependence of protein dynamics", *PROTEINS: Structure, Function, and Genetics*, (2001), **42**, 414–421.

[421] Wambolt CL and Saavedra SS, "Iodide fluorescence quenching of sol-gel immobilized BSA", *J. Sol-Gel Sci. and Technol.*, (1996), **1**, 53–57.

[422] Wan Q and Wang TH, "Single-crystalline Sb-doped SnO$_2$ nanowires: Synthesis and gas sensor application", *Chem. Comm.*, (2005), **30**, 3841–3843.

[423] Wan Q, Li QH, Chen YJ, Wang TH, Li XL, Gao G and Li JP, "Positive temperature coefficient resistance and humidity sensing properties of Cd-doped ZnO nanowires", *Appl. Phys. Lett.*, (2004), **84**, 3085–3087.

[424] Wan Q, Lin CL, Yu XB and Wang TH, "Room-temperature hydrogen storage characteristics of ZnO nanowires", *Appl. Phys. Lett.*, (2004), **84**, 124–126.

[425] Wang M, Singh H, Hatton TA and Rutledge GC, "Field-responsive superparamagnetic composite nanofibers by electrospinning", *Polymer*, (2004), **45**, 5505–5514.

[426] Wang R, Narang U, Prasad PN and Bright FV, "Affinity of antifluorescein antibodies encapsulated within a transparent sol-gel glass", *Anal. Chem.*, (1993), **65**, 2671–2675.

[427] Wang ZL, "Functional oxide nanobelts: Materials, properties and potential applications in nanosystems and biotechnology", *Annu. Rev. Phys. Chem.*, (2004), **55**, 159–196.

[428] Wang ZL and Song, "Piezoelectric nanogenerators based on zinc oxide nanowire arrays", *J. Science,* (2006), **312**, 242–246.

[429] Wang C, Chu X and Wu, M, "Detection of H2S down to ppb levels at room temperature using sensors based on ZnO nanorods", *Sens. Actuators B*, (2006), **113**, 320–323.

[430] Wang HT, Kang BS, Ren F, Tien LC, Sadik PW, Norton DP, Pearton SJ and Lin J, "Hydrogen-selective sensing at room temperature with ZnO nanorods", *Appl. Phys. Lett.*, (2005), **86**, 243503–243505.

[431] Wang Y, Jiang X and Xia YJ, "A solution-phase, precursor route to polycrystalline SnO2 nanowires that can be used for gas sensing under ambient conditions", *Am. Chem. Soc.*, (2003), **125**, 16176–16177

[432] Wang ZL, "Nanobelts, nanowires, and nanodiskettes of semiconducting oxides — from materials to nanodevices", *Adv. Mater.*, (2003), **15**, 432–436.

[433] Wang ZL, "Nanostructures of zinc oxide", *Mater. Today*, (2004), **6**, 26–33.

[434] Wegner K and Pratsinis SE, "Nozzle-quenching process for controlled flame synthesis of titania nanoparticles", *AIChE J.*, (2003), **49**, 1667–1675.

[435] Wei Y, Dong H, Xu J and Feng Q, "Simultaneous immobilization of horseradish peroxidase and glucose oxidase in mesoporous sol-gel host materials", *Chem. Phys. Chem.* (2002), **9**, 802–808.

[436] Wen X, Fang Y, Pang Q, Yang C, Wang J, Ge W, Wong KS and Yang S, "ZnO nanobelt arrays grown directly from and on zinc substrates: synthesis, characterization, and applications", *J. Phys. Chem.*, (2005), **109**, 15303–15308.

[437] Wheeler KE, Lees NS, Gurbiel RJ, Hatch SL, Nocek JM and Hoffman BM, "Electrostatic influence on rotational mobilities of sol-gel encapsulated solutes by nmr and epr spectroscopies", *J. Am Chem. Soc.*, (2004), **126**, 13459–13463.

[438] Wnek GE, *et al.*, "Electrospinning of nanofiber fibrinogen structures", *Nano Letters,* (2003), **3**, 213–216.

[439] Wu S, Ellerby LM, Cohan JS, Dunn B, El-Sayed MA and Valentine JS, "Bacteri-orhodopsin encapsulated in transparent sol-gel glass: A new biomaterial", *Zink JI Chem. Mater.*, (1993), **5**, 115–120.

[440] Wu N, Zhao M, Zheng JG, Jiang C, Myers B, Li S, Chyu M and Mao SX, "Porous CuO-ZnO nanocomposite for sensing electrode of high-temperature CO solid-state electrochemical sensor", *Nanotechnology*, (2005), **16**, 2878–2881.

[441] www.figaro.co.jp/en.company3.html

[442] Xia Y, Yang P, Sun Y, Wu Y, Mayers B, Gates B, Yin Y, Kim F and Yan H, "One-dimensional nanostructures: Synthesis, characterization and applications", *Advanced Materials*, (2003), **15**, 353–389.

[443] Xie J, Li X and Xia Y, "Putting electrospun nanofibers to work for biomedical research", *Macromolecular Rapid Communications*, (2008), **29**(22), 1775–1792.

[444] Xu JQ, Chen YP, Chen YD and Shen JN, "Hydrothermal synthesis and gas sensing characters of ZnO nanorods", *Sens. Actuators B.*, (2006), **113**, 526–531.

[445] Xu JQ, Chen YP, Li YD and Shen JN, "Gas sensing properties of ZnO nanorods prepared by hydrothermal method", *J. Mater. Sci.*, (2005), **40**, 2919–2921.

[446] Yamanaka SA, Nishida F, Ellerby LM, Nishida CR, Dunn B, Valentine JS and Zink JI, "Enzymatic activity of glucose oxidase encapsulated in transparent glass by the sol-gel method", *Chemistry of Materials*, (1992), 495–496.

[447] Yamazoe N, "New approaches for improving semiconductor gas sensors", *Sens. and Actuators B*, (1991), **5**, 7–19.

[448] Yang S, Iglesia E and Bell AT, "Oxidative dehydrogenation of propane over $V_2O_5/MoO_3/Al_2O_3$ and $V_2O_5/Cr_2O_3/Al_2O_3$: Structural characterization and catalytic function", *J. Phys. Chem.*, (2005), **109**, 8987–9000.

[449] Yang X, Li Z, Liu B, Klein-Hofmann A, Tian G, Feng Y, Ding Y, Su D and Xiao F, "'Fish-in-Net' encapsulation of enzymes in macroporous cages as stable, reusable, and active heterogeneous biocatalysts", *Adv. Mater.*, (2006), **18**, 410–414.

[450] Yang XG, et al., "Nanofibers of CeO_2 via an electrospinning technique", *Thin Solid Films*, (2005), **478**, 228–231.

[451] Yang XH, et al., "Preparation and characterization of ZnO nanofibers by using electr-lospun PVA/zinc acetate composite fiber as precursor", *Inorganic Chemistry Communications*, (2004), **7**, 176–178.

[452] Yao JL, Hao S and Wilkinson JS, "Indium tin oxide films by sequential evaporation", *Thin Solid Films*, (1990), **189**, 227–233.

[453] Yarin AL, Koombhongse S and Reneker DH, "Bending instability in electrospinning of nanofibers", *Journal of Applied Physics*, (2001), **89**, 3018–3026.

[454] Yarin AL, Koombhongse S, Reneker DH, "Taylor cone and jetting from liquid droplets in electrospinning of nanofibers", *Journal of Applied Physics*, (2001), **90**, 4836–4846.

[455] Yaszemski MJ, Oldham JB, Lu L and Currier BL, "Clinical needs for bone tissue engineering technology", in Davis JE, *Bone Engineering*, em squared, Toronto, (2000), 541–547.

[456] Ying Z, Wan Q, Song ZT and Feng SL, "Controlled synthesis of branched SnO_2 nanowhiskers", *Mater. Lett.*, (2005), **59**, 1670–1672.

[457] Ying Z, Wan Q, Song ZT and Feng SL, "SnO$_2$ nanowhiskers and their ethanol sensing characteristics", *Nanotechnology*, (2004), **15**, 1682–1684.

[458] Yongsunthon R and Lower SK, "Force spectroscopy of bonds that form between a staphylococcus bacterium and silica or polystyrene substrates", *Journal of Electron Spectroscopy and Related Phenomena*, (2006), **150**, 228–234.

[459] Yu J and Ju H, "Preparation of porous titania sol-gel matrix for immobilization of horseradish peroxidase by a vapor deposition method", *Anal. Chem.*, (2002), **74**, 3579–3583.

[460] Yu J and Ju H, "Amperometric biosensor for hydrogen peroxide based in hemoglobin entrapped in titania sol-gel film", *Analytica Chimia Acta*, (2003), **486**, 209–216.

[461] Yu J, Liu S and Ju H, "Mediator-free phenol sensor based on titania sol-gel encapsulation matrix for immobilization of tyrosinase by a vapor deposition method", *Biosensors and Bioelectronics*, (2003), **19**, 509–514.

[462] Yu HY, Kang BH, Pi UH, Park CW and Choi SY, "V$_2$O$_5$ nanowire-based nanoelectronic devices for helium detection", *Appl. Phys. Lett.*, (2005), **86**, 253102-1–253102-3.

[463] Yu, N., *et al.*, "Nanofibers of LiMn$_2$O$_4$ by electrospinning", *Journal of Colloid and Interface Science*, (2005), **285**(1), 163–166.

[464] Zarkoob S, *et al.*, *Structure and Morphology of Electrospun Silk Nanofibers*, Elsevier Sci Ltd., (2004).

[465] Zhang T, Tian B, Kong J, Yang P and Liu B, "A sensitive mediator-free tyrosinase biosensor based on an inorganic-organic hybrid titania sol-gel matrix", *Analytica Chimica Acta*, (2003), **489**, 199–206.

[466] Zhang D, Li C, Liu X, Han S, Tang T and Zhou C, "Detection of NO$_2$ down to ppb levels using individual and multiple In$_2$O$_3$ nanowire devices", *Appl. Phys. Lett.* (2003), **83**, 1845–1847.

[467] Zhang D, Liu Z, Li C, Tang T, Liu X, Han S, Lei B and Zhou C, "Detection of NO$_2$ down to ppb levels using individual and multiple In$_2$O$_3$ nanowire devices", *Nano Lett.*, (2004), **4**, 1919–1924.

[468] Zhang G, *et al.*, *Electrospun Nanofibers for Potential Space-based Applications*, Elsevier Science, (2005).

[469] Zhang N, Nichols HL, Tylor S and Wen X, "Fabrication of nanocrystalline hydroxyapatite doped degradable composite hollow fiber for guided and biomimetic bone tissue engineering", *Mat. Sci. Eng. C-Bio. S*, (2006), **27**, 599.

[470] Zhang Y, Kolmakov A, Lilach Y and Moskovits, "Electronic control of chemistry and catalysis at the surface of an individual tin oxide nanowire", *M. J. Phys. Chem. B*, (2005), **109**, 1923–1929.

[471] Zheng L and Brennan JD, "Measurement of intrinsic fluorescence to probe the conformational flexibility and thermodynamic stability of a single tryptophan protein entrapped in a sol-gel derived glass matrix", *Analyst*, (1998), **123**, 1735–1744.

[472] Zheng L, Reid WR and Brennan JD, "Measurement of fluorescence from tryptophan to probe the environment and reaction kinetics within protein doped sol-gel derived glass monoliths", *Anal. Chem.*, (1997), **69**, 3940–3949.

[473] Zhou H, "Loops, linkage, rings, catenanes, cages, and crowders: Entropy based strategies for stabilizing proteins", *Acc. Chem. Res.*, (2004), **37**, 123–130.

[474] Zhou H and Dill KA, and references there in, "Stabilization of proteins in confined spaces", *Biochemistry,* (2001), **40**, 11289–11293.

[475] Zhou J, "Protein folding and binding in confined spaces and in crowded solutions", *J. Mol. Recognit.,* (2004), **17**, 368–375.

[476] Zhou JWL, Chan HY, To TKH, Lai KWC and Li WJ, "Polymer MEMS actuators for underwater micromanipulation", *IEEE/ASME Trans. on Mechatronics,* (2004), **9**, 334–342.

[477] Zhu BL, Xie CS, Wang AH, Zeng DW, Song WL and Zhao XZ, "The gas-sensing properties of thick film based on tetrapod-shaped ZnO nanopowders", *Mater. Lett.,* (2005), **59**, 1004–1007.

[478] Zolkov C, Avnir D and Armon R, "Tissue-derived cell growth on hybrid sol-gel films", *J. Mater. Chem.,* (2004), **14**, 2200–2205.

[479] Zong XH, *et al.,* "Structure and process relationship of electrospun bioabsorbable nanofiber membranes", *Polymer,* (2002), **43**, 4403–4412.

Color Index

Color Index

CHAPTER 2

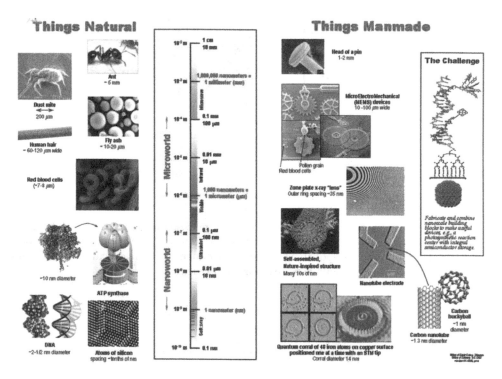

Figure 2.1 The Scale of things. [From www.nano.gov/html/facts/The_Scale_of_Things.html, National Nanotechnology Initiative- NSF.]

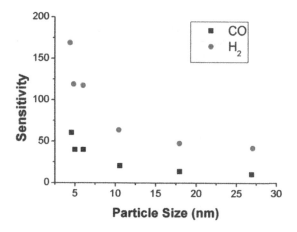

Figure 2.2 The effect of particle size on the gas sensitivity for a SnO_2 sensor exposed to CO and H_2. [Reprinted from Sensors and Actuators B, 3(2) pp. 147–155, C. Xu, J. Tamaki, N. Miura, and N. Yamazoe, Grain size effects on gas sensitivity of porous SnO2-based elements © (1991), with permission from Elsevier.]

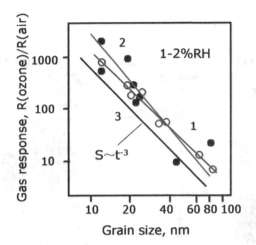

Figure 2.3 Influence of In_2O_3 film grain size on the gas response to ozone. [Reprinted from Thin Solid Films, 460, G. Korotcenkov, V. Brinzari, A. Cerneavschi, M. Ivanov, V. Golvanov, A. Cornet, J. Morante, A. Cabot and J. Arbiol, The influence of film structure on In_2O_3 gas response, p. 315, © (2004), with permission from Elsevier.]

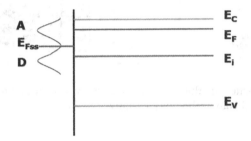

Figure 2.4 'Flat Band' condition-No charge exchange between the surface states and the bulk. [Figures 2.4–2.7 Reprinted from Semiconductor Sensors, S. M. Sze, 1st Edition, 1994 (John Wiley & Sons Inc.), with permission from John Wiley & Sons Inc.]

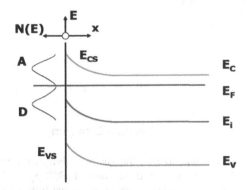

Figure 2.5 'Band Bending'-where the electrons from the semiconductor surface have moved to the surface states[3].

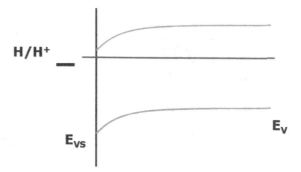

Figure 2.6 Formation of an accumulation layer between the electropositive surface species and the negatively charged semiconductor.

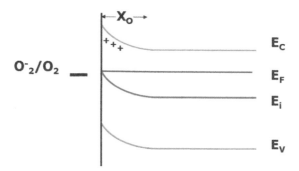

Figure 2.7 Formation of a depletion layer between the negatively charged surface species and the positive donor ions.

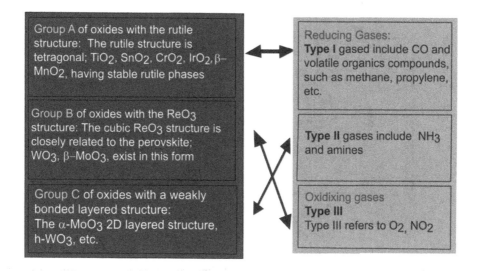

Figure 2.8 Semi-empirical map of Gas-Oxide Interactions. [From Gouma, *Rev. Adv. Mater. Sci.*, 5, pp. 123–138, 2003.]

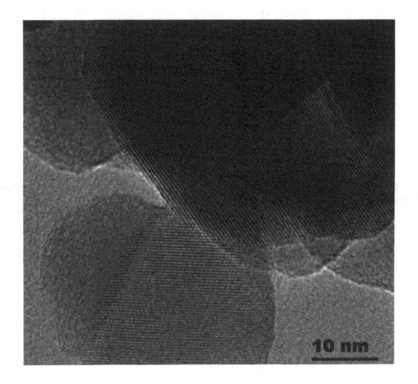

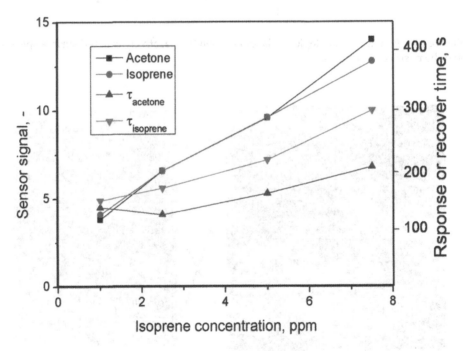

Figure 2.13 **(top)** HRTEM image of nanocrystalline flame-sprayed titania; **(bottom)** comparative plot of sensing data for isoprene and ethanol.

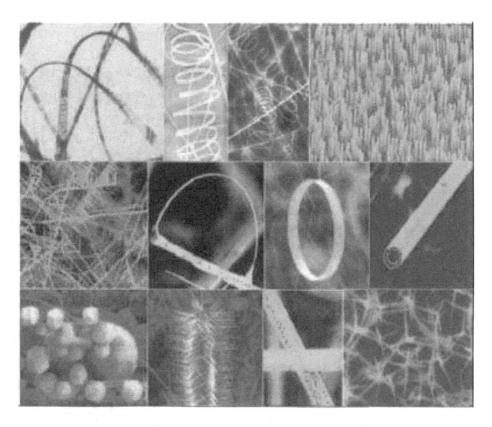

Figure 2.15 ZnO nanostructures synthesized under controlled conditions by thermal evaporation of solid powders. [Reprinted from Materials Today, 6, Z. L. Wang, Nanostructures of Zinc Oxides, p.26, © (2004), with permission from Elsevier.]

CHAPTER 3

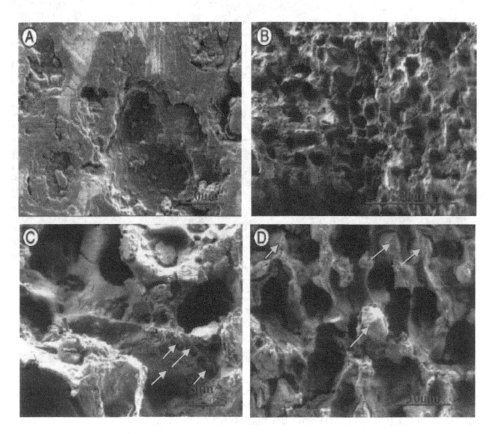

Figure 3.2 Scanning electron micrographs of *Bacillus Sphaericus* or *Saccharomyces Cerevisiae* (yeast) encapsulated within a ceramic by the freeze gelation method. A slurry of ceramic powders and biological component is solidified at freezing at very low temperatures. Thus-formed biodoped ceramic retains its shape even when the temperature becomes normal. (A) Surface of a *B. Sphaericus*-encapsulating ceramic. (B and C) Fracture surfaces of *B. Sphaericus* encapsulating ceramic. (D) Fracture surface of a *S. Cerevisiae*-encapsulating ceramic. Embedded cells are marked with arrows (Soltmann *et al.*, 2003).

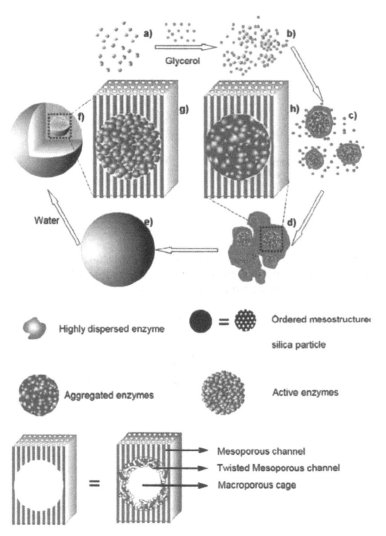

Figure 3.3 "Fish-in-Net" encapsulation of enzymes in macroporous cages as stable, reusable, and active heterogeneous biocatalysts. (A) Buffer solution containing highly dispersed enzymes. (B) Preformed precursor with ordered mesostructured silica particles mixed with enzyme. Tetraethyl orthosilicate was assembled from a triblock ethylene oxide(EO)/propylene oxide (PO) co-polymer surfactant ($EO_{20}PO_{70}EO_{20}$,P123) in ethanol. Evaporation of ethanol and addition of glycerol yielded these preformed precursors. (C) Assembly of enzymes with ordered mesostructured silica particles. (D) Interactions between enzyme and preformed precursor led to tailoring of pores. Enzymes were trapped in the "net" formed by polymerization and condensation of ordered mesostructured silica particles. (E) Formation of silica spheres encapsulating enzymes in microporous cages. (F) Silica spheres with encapsulated enzyme under biocatalytic conditions. When water is introduced into these cages, and the chemical environment is similar to that of a native enzyme in solution, it is possible that enzymes are dispersed. (G) Magnification of the enzyme-filled macroporous cage from (F). (H) Magnification of the enzyme-filled macroporous cage from (D) where enzymes are aggregated in macroporous cages. [Reprinted from Yang *et al.*, 2006, with permission from WILEY-VCH Verlag GmbH & Co.]

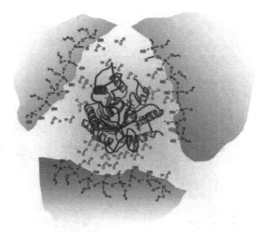

Figure 3.6 Schematic view of the entrapped enzyme with a few water molecules inside a pore, two of which are protonated; the nominal pH is very low. [Reprinted from Frenkel-Mullerad and Avnir, 2005, with permission from American Chemical Society.]

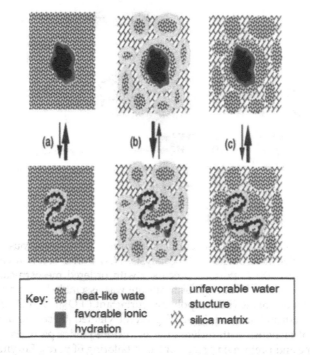

Figure 3.8 Role of bulk water structure in protein folding equilibria and the hydrophobic effect. A hypothetical two-state equilibrium is depicted for a folded globular protein (top panels) and the unfolded random-coil conformation (lower panel) in three different aqueous environments. (a) In an ideal dilute solution, the folded conformation is favored due to a strong hydrophobic effect. (b) The glass-entrapped protein is influenced by the unfavorable water structure at the silica interface of the surrounding pores. The unfolded state predominates due to a diminished hydrophobic effect. (c) Addition of compatible solutes reduces the average free energy of the bulk water to a value that more closely resembles neat water. The native state is favored due to the restored driving force of the hydrophobic effect. [Reprinted from Eggers and Valentine, 2001, with permission from Elsevier.]

CHAPTER 4

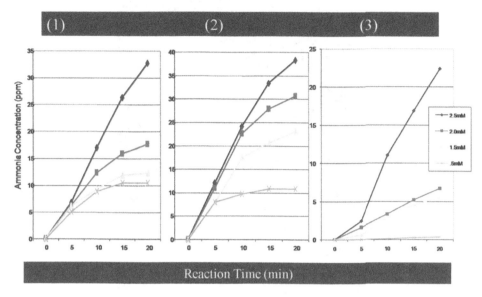

Figure 4.4 Ammonia concentration vs. time when urea solutions reacted with (1) 0.2 ml of urease in PBS buffer, (2) 0.2 ml 30% urease in buffer/70% PVP in ethanol solution, and (3) 0.1 ml of urease/PVP nanofiber mat.

CHAPTER 5

Figure 5.2 SACMI's E-nose is based on 6 SnO_2 sensors.

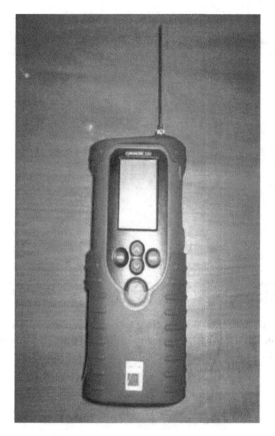

Figure 5.3 The Cyranose E-nose based on polymer sensor arrays.

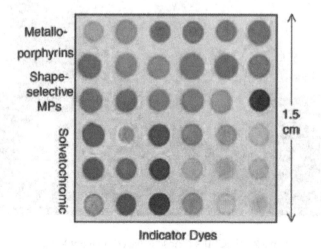

Figure 5.4 Suslick's colorimetric sensor aray.

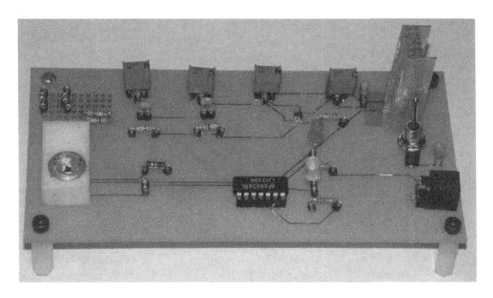

Figure 5.7 Acetone breathalyzer prototype.

Figure 5.17 Chemo-mechanical actuation of PANI-CA hybrid films in acetone vapor.

Index